McGraw-Hill
netw⊙rk
A Social Studies Learning System

Discovering World Geography

READING ESSENTIALS & STUDY GUIDE

Student Workbook

Mc
Graw
Hill
Education

Bothell, WA • Chicago, IL • Columbus, OH • New York, NY

www.mheonline.com/networks

ISBN: 978-0-07-663608-2
MHID: 0-07-663608-9

Printed in the United States of America.

6 7 8 9 RHR 18 17 16

Table of Contents

Table of Contents *continued*

Table of Contents *continued*

Dear Student,

We know that taking notes, using graphic organizers, and developing critical thinking skills are vital to achieving academic success. Organizing solid study materials can be an overwhelming task. McGraw-Hill has developed this workbook to help you master content and develop the study skills necessary for academic success.

This workbook includes all of the core content found in the *Discovering World Geography* program. The note-taking and graphic organizer activities will help you learn to organize content for improved comprehension and testing.

Note-Taking System

You will notice that the pages in the *Reading Essentials and Study Guide* are arranged in two columns. The wide column contains text and graphics that summarize each lesson of the chapter. The questions in the narrow column will help you focus on the main concepts and help you improve your note-taking skills.

Graphic Organizers

Many graphic organizers appear in this workbook. Graphic organizers allow you to see the lesson's important information in a visual format. In addition, they help you summarize information and remember the content.

networks

The Geographer's World

Lesson 1: How Geographers View the World

ESSENTIAL QUESTION
How does geography influence the way people live?

Terms to Know

geography the study of Earth and its people, places, and environments
spatial Earth's features in terms of their places, shapes, and relationships to one another
landscape the portions of Earth's surface that can be viewed at one time from another location
relative location the location of one place compared to another place
absolute location the exact location of something
latitude the lines on a map that run east to west
Equator a line of latitude that runs around the middle of Earth
longitude the lines on a map that run north to south
Prime Meridian the starting point for measuring longitude
region a group of places that are close to one another and share some characteristics
environment the natural surroundings of a place
landform the shape and nature of the land
climate the average weather in an area over a long period of time
resource a material that can be used to produce crops or other products

Where in the world?

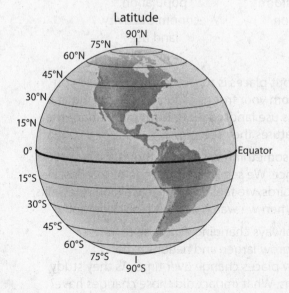

Latitude

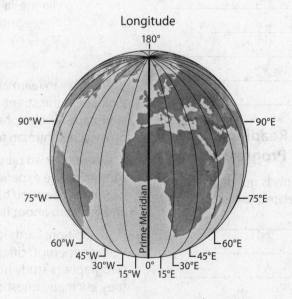

Longitude

networks

The Geographer's World

Lesson 1: How Geographers View the World, *Continued*

Marking the Text

1. Read the text on the right. Highlight the different ways it means to think spatially.

Defining

2. Give three additional examples of landscapes.

_____

_____

_____

Making Connections

3. Describe how you experience geography in your daily life.

_____

_____

_____

_____

_____

_____

Reading Progress Check

4. How is geography related to history?

_____

_____

_____

_____

Geographers Think Spatially

Guiding Question *What does it mean to think like a geographer?*

Geographers try to understand the world. They look at people and the world in which they live. Geographers study why people live where they do. They also study how people relate to each other and to their environment. **Geography** is the study of Earth and its people, places, and environments.

One way geographers look at Earth is spatially. **Spatial** means "taking up space." Therefore, they study the things that take up space on Earth, including cities, countries, mountains, and lakes.

Thinking spatially means that you consider:

- where things are located on a map,
- their size and direction,
- how far apart they are,
- how different places are related to each other,
- and what characteristics they have.

Earth is filled with both physical and human features.

Physical Features

climate, landforms, vegetation

Human Features

population, economic activity, land use

One way to learn about places is to study **landscapes**. If you look out at the street from your front porch, you are looking at a landscape. Geographers use landscapes to learn more about the physical and human features they see.

Geography isn't just something we look at, though. It is something we experience. We see it in the change of seasons, we hear it in the sound of birds, we feel it when we walk on sidewalks, and we learn about it when we watch TV. This is all geography.

In addition, Earth is always changing. Rivers shift course, volcanoes erupt, cities grow larger, and nations expand. Geographers study how places change over time. As they study, they ask many questions. What impact did those changes have? What factors made a city grow? What effect did its growth have on the people who lived there?

networks

The Geographer's World

Lesson 1: How Geographers View the World, *Continued*

The Five Themes of Geography

Guiding Question *How can you make sense of a subject as large as Earth and its people?*

Geographers like to organize information about the world into five themes: location, place, region, human interaction with the environment, and movement.

The first theme is location. Location is where something is found on Earth. There are two types of location: relative and absolute. **Relative location** describes where a place is compared to another place. For example, knowing that New Orleans is near the mouth of the Mississippi River helps us understand why the city became an important trading port.

Absolute location is the *exact* location of something. Maps show absolute location with horizontal and vertical lines. Horizontal lines are called lines of **latitude**. They run east to west and measure distance going north to south. The **Equator** is a line of latitude in the middle of Earth and has latitude of 0°.

Vertical lines are called lines of **longitude**. They run north to south and measure distance going east to west. The **Prime Meridian** is the starting point for measuring longitude. It has a longitude of 0°.

Place and region are additional themes. When places are close to one another, they share some features, which means they belong to the same **region.** Both places and regions can be described by human and physical features. Geographers like to study regions so they can look for patterns in larger areas. They also want to examine special features that make each place in a region unique.

The fourth theme is how humans interact with their environment. The **environment** is the natural surroundings of a place. It includes landforms, climate, and resources. **Landforms** are the shape and nature of the land. Hills, valleys, and mountains are all types of landforms. **Climate** is the average weather in a place over a long period of time. **Resources** are the materials that come from a place. They are used to plant crops or make other products.

The final theme is movement. People might have to move for many reasons, such as war, famine, or religious freedom. When a lot of people move, there might be problems with not enough housing or jobs. Products also move from place to place. Roads, ships, airplanes and trucks are important to movement. Ideas can move over the Internet, telephone, television, and radio.

? Locating

5. Describe the difference between relative and absolute location.

_____

_____

_____

_____

_____

✓ Reading Progress Check

6. How is the theme of location related to the theme of place?

_____

_____

_____

_____

_____

? Summarizing

7. Summarize how people, products, ideas, and information move from place to place.

_____

_____

_____

_____

_____

_____

_____

The Geographer's World

Lesson 1: How Geographers View the World, *Continued*

✓ Reading Progress Check

8. How do geographers use visuals?

_____

_____

_____

_____

⚙ Drawing Conclusions

9. Why is it important for geographers to ask *why* and *how* questions?

_____

_____

_____

Skill Building

Guiding Question *How will studying geography help you develop skills for everyday life?*

Geographers use maps and many other visuals to show a representation of the world. Visuals include graphs, charts, diagrams, and photographs. By studying geography, you can learn to use these visuals, too.

Visual	Use
Graphs	Help compare information
Charts	Show information in columns and rows
Diagrams	Use pictures to show something in the world

Geographers also ask analytical questions about the world. They may want to examine causes and effects. Or they may want to analyze how something changed over time. For instance, they may ask: How does climate affect how people live? Or they may ask: Why do people in different nations use resources differently?

As you study geography, you can learn to ask—and answer—questions like these, too.

Writing

Check for Understanding

1. **Narrative** Describe your neighborhood spatially. Include its location, shape, characteristics, and the relationship between human and physical features.

_____

_____

2. **Informative/Explanatory** Give an example of how humans can affect the environment and how the environment can affect humans.

_____

_____

_____

_____

The Geographer's World

Lesson 2: The Geographer's Tools

ESSENTIAL QUESTION

How does geography influence the way people live?

Terms to Know

hemisphere each half of Earth
key the feature on a map that explains the symbols, colors, and lines on a map
scale bar the feature on a map that tells how a measured space on the map relates to actual distances on Earth
compass rose the feature on a map that shows direction
map projection the system of presenting the round Earth on a flat map
scale the relationship between distances on the map and on Earth
elevation the height or depth of a place compared to sea level
relief the difference between the elevation of one feature and the elevation of another feature near it
thematic map a map that shows more specialized information
technology any way that scientific discoveries are applied to practical use
remote sensing getting information from far away

Where in the world?

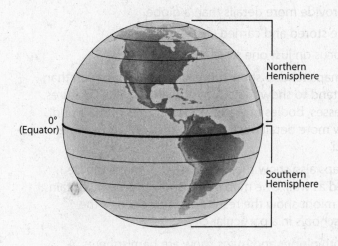
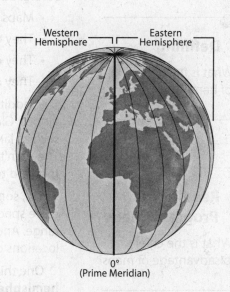

The Geographer's World

Lesson 2: The Geographer's Tools, *Continued*

Marking the Text

1. Read the text on the right. Highlight the advantages of globes in one color, and the advantages of maps in another color.

Comparing and Contrasting

2. Why is it possible for maps to show more detail than globes?

_____

_____

_____

_____

_____

Defining

3. What is the definition of hemisphere?

_____

_____

_____

Reading Progress Check

4. What is the chief disadvantage of maps?

_____

_____

_____

Using Globes and Maps

Guiding Question *What is the difference between globes and maps?*

Making and using maps and globes is a big part of geography. Geographers use both of them to picture the world and show where things are located.

A globe is the most accurate way to show places on Earth. Because Earth is in the shape of a sphere, globes are also in the shape of a sphere. A sphere is another word for *ball*.

Globes have some advantages over maps.

- Globes are shaped like Earth.
- They show the correct shapes of land and bodies of water.
- They show accurate distances and directions between places on Earth.

Maps are not round like globes. Instead, they are flat representations of the round Earth. Because of this, maps distort what Earth looks like. *Distort* means "bend or twist." In other words, maps will always show the physical features of Earth incorrectly to some degree.

Even though maps aren't as accurate as globes, they have several advantages.

- Maps don't have to show the whole planet.
- They can provide more details than a globe.
- They can be stored and carried easily.
- They can focus on just one small area.

In addition, maps tend to show more kinds of information than globes. Globes tend to show major physical and political features, such as land masses, bodies of water, and countries of the world. They can't show more detail because then they would become too hard to read.

Yes, some maps also show these features. But maps can be more specialized as well. One map might show a large mountain range. Another might show the results of an election or the locations of all schools in a particular city.

One thing both globes and maps show are hemispheres. A **hemisphere** is half of Earth. These halves are divided by the Equator and the Prime Meridian. The Equator divides Earth into the Northern and Southern Hemispheres. The Prime Meridian and the International Date Line divide Earth into the Eastern and Western Hemispheres.

netw⦿rks

The Geographer's World

Lesson 2: The Geographer's Tools, *Continued*

All About Maps

Guiding Question *How do maps work?*

Maps are everywhere—in the subway, in your textbook, at a Web site, or in a state park. Each map is different, but they share many common features. The table below describes these features.

Feature	Description
Title	Tells what area the map will cover
Key	Unlocks the meaning of the map by explaining symbols, colors, and lines
Scale bar	Tells how a measured space on the map corresponds to actual distances on Earth
Compass Rose	Shows north, south, east, and west direction

But how is a round Earth shown on a flat map? To do this, geographers use something called a **map projection**. This is when some parts of Earth are distorted in order to represent other parts as accurately as possible. Some projections show the correct size of certain areas in relation to one another. Others break apart the oceans. Mapmakers choose a projection based on the purpose of the map.

As shown in the table, the scale bar tells how distance is measured on a map and compares it to actual distances on Earth. A large-scale map focuses on a small area. Its **scale** might be one inch to 10 miles (16 km). This means that one inch on the map equals 10 miles (16 km) on Earth. A small-scale map focuses on a larger area and has a larger scale.

There are two basic types of maps: general purpose and thematic. General purpose maps show either human or physical features. A political map shows human features, such as the boundaries of a country.

A physical map shows natural features, such as mountains and rivers. Many physical maps also display **elevation**, or how much above sea level something is. Physical maps might also show **relief**, or the difference between the elevation of one feature and the elevation of another feature.

Thematic maps show more specialized information. For instance, a thematic map might display the kinds of plants that grow in different areas. Or it might show where ranching or mining take place.

Explaining

5. In your own words, describe a map projection.

_____

_____

_____

_____

_____

Reading Progress Check

6. How do the two main types of maps differ?

_____

_____

_____

_____

_____

Defining

7. What is the difference between elevation and relief?

_____

_____

_____

_____

_____

networks

The Geographer's World

Lesson 2: The Geographer's Tools, *Continued*

⚙️ **Sequencing**

8. Highlight the text that explains how GPS systems work. List the steps involved in calculating a location on earth.

_____

_____

_____

✅ **Reading Progress Check**

9. How could remote sensing be used as part of a GIS?

_____

_____

Geospatial Technologies

Guiding Question *How do geographers use geospatial technologies?*

Cell phones and GPS devices use something called geospatial technology. **Technology** is any way that scientific discoveries are applied to practical use. Geospatial technologies help us think spatially about geography. GPS devices are one type of geospatial technology. They work with something called the Global Positioning System (GPS).

The GPS system is made up of more than 30 satellites that orbit Earth. These satellites send out radio signals to GPS devices on Earth. GPS devices receive the signals from four satellites. They need four satellites so they can combine the signals. This helps them calculate their exact location on Earth.

Another type of geospatial technology is the geographic information system (GIS). This system gathers, stores, and analyzes geographic information. The information is then shown on a computer screen as a map.

Satellites gather information by something called **remote sensing**. This means the information comes from far away. Satellites might take pictures of the land. Or they might measure the amount of moisture in the soil. In the 2000s, GIS technology uses remote sensing to help protect plants and animals.

Writing

Check for Understanding

1. **Informative/Explanatory** Why do maps distort the way Earth's surface really looks?

_____

_____

_____

2. **Informative/Explanatory** Are road maps general-purpose maps or thematic maps? Explain your answer.

_____

_____

_____

networks

Physical Geography

Lesson 1: Planet Earth

ESSENTIAL QUESTION
How does geography influence the way people live?

Terms to Know
orbit to circle around something
axis an imaginary line that runs through the Earth's center from the North Pole to the South Pole
revolution a complete trip of Earth around the sun
atmosphere the layer of gases that surrounds Earth
solstice one of two days each year when the sun reaches its northernmost or southernmost point
equinox one of two days each year when the sun is directly overhead at the Equator
climate the average weather conditions in an area over a long period of time
precipitation the water that falls on Earth as rain, snow, sleet, hail, or mist
rain shadow an area that receives reduced rainfall because it is on the side of a mountain facing away from the ocean

Where in the World: Tectonic Plate Boundaries

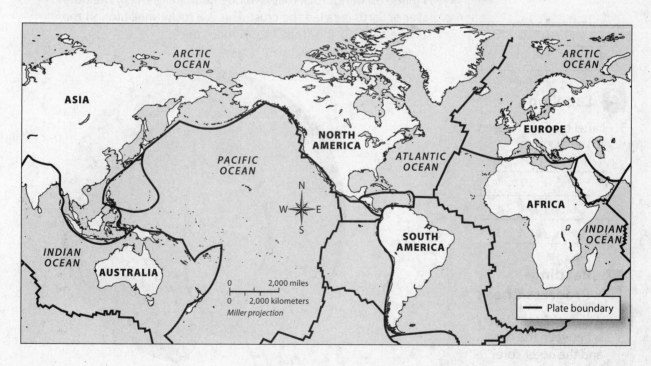

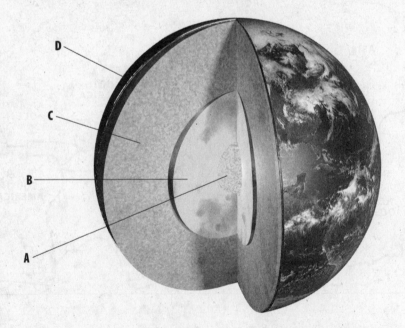

networks

Physical Geography

Lesson 1: Planet Earth, *Continued*

Defining

1. What is the difference between *rotation* and *revolution*?

   _____

   _____

   _____

   _____

   _____

Explaining

2. What is the purpose of adding an extra day to the calendar every four years?.

   _____

   _____

   _____

   _____

Labeling

3. Label the Earth's layers in the diagram.

   _____

   _____

   _____

   _____

Reading Progress Check

4. Which of Earth's layers is between the crust and the outer core?

   _____

Looking at Earth

Guiding Question *What is the structure of Earth?*

The sun is at the center of our solar system. Earth and other planets **orbit**, or move around, the sun. Heat and light from the sun make life on Earth possible.

Earth spins on its axis. The axis is an imaginary line that runs through Earth's center from the North Pole to the South Pole. Earth spins once around its **axis** every 24 hours. This is called one rotation. This rotation puts some areas of Earth in sunlight and some in darkness. These periods are day and night.

Earth makes one complete trip around the sun each year. This is called one **revolution**. A complete revolution takes 365 ¼ days. Most calendar years are 365 days, but one day is added to the end of February every fourth year. This extra day is called a leap day and these years are called leap years.

Earth is made of three layers. The center of Earth is called the core. The core actually has two parts: a solid inner core and a liquid metal outer core. The mantle is the middle layer. The mantle is made up of hot rock that is about 1,800 miles thick. The outer layer of Earth is called the crust. This is a rocky shell around the surface of Earth. It is from 2 to 75 miles thick.

D

C

B

A

Earth has four major physical subsystems: the hydrosphere, the lithosphere, the atmosphere, and the biosphere.

Physical Geography

Lesson 1: Planet Earth, *Continued*

The hydrosphere contains all of Earth's water. About 70 percent of Earth's surface is covered in water. This water is in oceans, seas, lakes, ponds, rivers, groundwater, and ice.

The lithosphere is Earth's land. The land includes mountains, hills, plateaus, plains, and the land beneath the ocean.

The **atmosphere** is the layer of gases that surrounds Earth. Most of Earth's atmosphere is contained within 16 miles of Earth's surface. Outer space begins 100 miles above the surface of Earth.

The biosphere includes all living things on Earth or in the atmosphere. Animals, plants, and people are part of the biosphere.

Earth's Physical Subsystems	
hydrosphere	Earth's water
lithosphere	Earth's land
atmosphere	gases surrounding Earth
biosphere	living things on Earth

Seasons

Guiding Question *How does Earth's orbit around the sun cause seasons?*

Earth is tilted. Because of the tilt, not all places on Earth receive the same amount of direct sunlight at the same time. About half of Earth is tilted toward the sun and half is tilted away from the sun at any time. Because of this, when it is winter in the Northern Hemispheres it is summer in the Southern Hemisphere.

The North Pole is tilted furthest toward the sun on about June 21. In the Northern Hemisphere this is called the summer **solstice**, or the beginning of summer. The Northern Hemisphere receives the most direct rays of the sun at this time of year and has the most hours of daylight each day.

The North Pole is tilted furthest away from the sun on about December 22. In the Northern Hemisphere this is the winter solstice, the day on which winter begins. The sun's rays reach their southernmost point on this day. The Southern Hemisphere experiences summer and long hours of daylight at this time.

The fall **equinox** is about September 23 and the spring equinox is about March 21. On an equinox, the sun's rays are directly overhead at the Equator. Day and night in both hemispheres are about 12 hours long everywhere on Earth on an equinox.

Marking the Text

5. Circle the physical subsystem that includes the air that we breathe. Underline the physical subsystem that includes humans. Double underline the physical subsystem that includes the North American continent.

? Taking Notes

6. Read the text to the left. Complete the chart below.

Spring equinox	
	June 21
	September 23
Winter solstice	

✔ Reading Progress Check

7. When it is winter in the Southern Hemisphere, what season is it in the Northern Hemisphere?

_____

_____

Physical Geography

Lesson 1: Planet Earth, *Continued*

? Explaining

8. Why are the tropics hot and sunny?

_____

_____

_____

⚙ Determining Cause and Effect

9. Which side of a mountain gets less precipitation? Explain.

_____

_____

_____

_____

_____

✏ Marking the Text

10. Underline each factor that influences climate.

✓ Reading Progress Check

11. Do the words *weather* and *climate* have the same meaning? Explain.

_____

_____

_____

_____

_____

Factors That Influence Climate

Guiding Question *How do elevation, wind and ocean currents, weather, and landforms influence climate?*

Near the Equator, the sun's rays are nearly directly overhead throughout the year. This area is known as the tropics. As a result, the climate in the tropics is hot and sunny. The sun's rays hit at a slant north and south of the tropics. The North and South Poles get no sunlight at all during certain months. These regions are cold throughout the year.

Elevation, or a place's distance above sea level, also influences climate. The air thins as altitude increases. Thinner air holds less heat.

Wind and water currents help circulate the sun's heat around the globe. Warm air rises. Cooler air rushes in under the warm air, causing wind. In the oceans, warm water moves away from the Equator while cold water flows toward it from the poles.

Weather and climate are different. Weather is the state of the atmosphere at any given time. It is easy to tell the weather by looking out the window or reading a thermometer. **Climate** is the average weather conditions in an area over a longer period. Average daily temperature is one measure used to compare climates. It is the average of the highest and lowest temperature in an area during a 24-hour period.

Climate also includes rainfall or snowfall that occurs year after year. Any water that falls on Earth is called **precipitation**. It includes rainfall, snowfall, hail, sleet, and mist. Measuring the amount of precipitation an area receives in a season or a year provides information about the area's climate.

Landforms like mountains also affect climate. Wind blows inland from oceans. As the air rises over mountains it cools and moisture condenses. This causes rain to fall on the side of the mountain facing the ocean. The other side of the mountain is in a **rain shadow**, where much less precipitation falls.

Different Types of Climate Zones

Guiding Question *What are the characteristics of Earth's climate zones?*

Earth is divided into six climate zones. These zones are classified by temperature, amount of precipitation, and the distance from the Equator.

Physical Geography

Lesson 1: Planet Earth, *Continued*

The Earth's Climate Zones	
Tropical	Hot and rainy; dense forests
Desert	Dry; hot or cold
Humid Temperate	Changing seasons and varying types of weather
Cold Temperate	Generally cold with a short summer
Polar	Very cold; covered in snow and ice
High Mountain	Variable conditions at the top of high mountain ranges

Each climate zone has different kinds of animal and plant life. These animals and plants are adapted to do well within that climate. A biome is a type of large ecosystem with similar life forms and climates. Earth has several biomes: rain forest, desert, grassland, and tundra.

Many scientists say that climates are changing around the world. Global warming is an increase in the average temperature of Earth's atmosphere. Many scientists say a buildup of pollution in the air is contributing to the rising temperature. If the temperature keeps rising, climates around the world will change. Plant and animal species will have difficulty adapting to these changes.

Drawing Conclusions

12. Which of the six major climate zones is likely to be the coldest? Which is likely to be the hottest? Explain.

_____

_____

_____

_____

Reading Progress Check

13. In which one of the six major climate zones do you live?

_____

_____

Writing

Check for Understanding

1. **Informative/Explanatory** Why does February 29th occur every four years?

_____

_____

_____

2. **Informative/Explanatory** In which climate zone would you most want to live? Explain.

_____

_____

_____

_____

networks

Physical Geography

Lesson 2: A Changing Earth

ESSENTIAL QUESTION
How does geography influence the way people live?

Terms to Know

continent a large, unbroken mass of land

tectonic plate one of the 16 pieces of Earth's crust

fault a place where two tectonic plates grind against each other

earthquake a shaking or trembling of Earth, caused by the collision of tectonic plates

Ring of Fire a long, narrow band of volcanoes surrounding the Pacific Ocean

tsunami a giant ocean wave caused by volcanic eruptions or movement of the earth under the ocean floor

weathering the process by which Earth's surface is worn away by natural forces

erosion the process by which weathered bits of rock are moved elsewhere by water, wind, or ice

glacier a large body of ice that moves slowly across land

When did it happen?

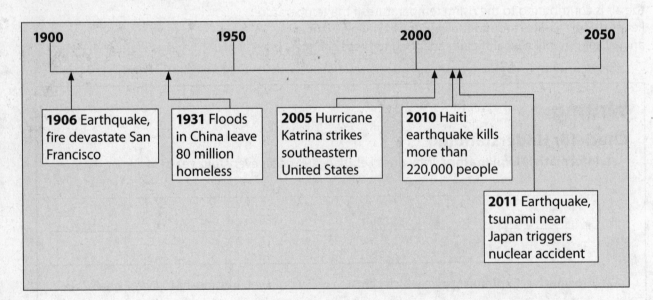

1900   1950   2000   2050

1906 Earthquake, fire devastate San Francisco

1931 Floods in China leave 80 million homeless

2005 Hurricane Katrina strikes southeastern United States

2010 Haiti earthquake kills more than 220,000 people

2011 Earthquake, tsunami near Japan triggers nuclear accident

networks

Physical Geography

Lesson 2: A Changing Earth, *Continued*

Forces of Change

Guiding Question *How was the surface of Earth formed?*

A **continent** is a large, unbroken mass of land. The continents have shifted and moved over time. Earth has seven unique continents:

- Asia
- Africa
- North America
- South America
- Europe
- Antarctica
- Australia

The surface of Earth changes over time. Even though you usually cannot feel it, the land beneath your feet is moving. This is because Earth's crust is not a solid sheet of rock. Earth's surface is more like many puzzle pieces pushed close together, floating on a sea of boiling rock. The movement of the pieces creates many of Earth's landforms.

Earth's rigid crust is made up of 16 enormous pieces called **tectonic plates**. These plates move at different speeds over the liquid mantle below them. Some of the plates move a few inches a year. It takes millions of years for plates to move enough to create new landforms.

When plates collide or move apart, new landforms can be made. When they hit, mountain ranges can result. When they move apart, a crack can form and magma can rise through the crust and form new crust as it cools.

Sometimes sudden changes occur. A **fault** is a place where two plates grind together. When rocks on either side of a fault move, an **earthquake**, a shaking or trembling of Earth's crust, can result. Earthquakes can also occur when volcanoes erupt.

Several tectonic plates lie at the bottom of the Pacific Ocean. Over time, the edges of these plates have been forced under the edges of plates surrounding the ocean. These movements led to the creation of a long, narrow band of volcanoes around the Pacific Ocean called the **Ring of Fire**.

An earthquake causes shaking of Earth. When this occurs in the ocean, it can cause an enormous wave called a **tsunami**. Tsunamis can cause terrible destruction along a coastline.

 Defining

1. Tectonic plates are parts of what layer of Earth?

   _____

✏️ **Marking the Text**

2. Circle the number of tectonic plates.

✏️ **Marking the Text**

3. Highlight the two causes of earthquakes.

⚙️ **Drawing Conclusions**

4. How did the Ring of Fire get its name?

   _____
   _____
   _____

✅ **Reading Progress Check**

5. Earth's surface plates are moving. Why don't we feel the ground moving under us?

   _____
   _____
   _____
   _____

Physical Geography

Lesson 2: A Changing Earth, *Continued*

Copyright © The McGraw-Hill Companies, Inc.

Defining

6. What is the difference between weathering and erosion?

_____
_____
_____
_____
_____

Making Inferences

7. What are two ways that an ocean can affect a coastline?

_____
_____
_____
_____
_____

Marking the Text

8. Masses of ice and snow are found in three forms. Underline the smallest mass. Double underline the next size mass. Triple underline the largest mass.

Other Forces at Work

Guiding Question *How can wind, water, and human actions change Earth's surface?*

Earth's surface is also changed by **weathering**. Weathering is the process by which Earth's surface is worn away by forces such as wind, rain, chemicals, and the movement of ice and flowing water. Plants can cause weathering too. Small seeds fall into cracks in rocks. When rain falls it causes seeds to sprout. As plant roots expand they can split the rock apart.

The effects of weathering may not be apparent from one day to the next. They only become obvious over time. Edges become chipped and worn. Raised surfaces become smooth. Rocks and soil break down and wear away. The Appalachian Mountains in the United States east of the Mississippi River have become worn down and rounded over millions of years of weathering.

Erosion is a process that works with weathering to change surface features of Earth. Erosion is a process by which weathered bits of rock are moved by water, wind, or ice. Fantastic formations can occur as a result of erosion. The Grand Canyon was formed by weathering and erosion. Water flowed over the region and weakened the rock. Bit by bit the moving water and wind took away small pieces of rock until it carved a deep, wide channel across the land.

Different types of rock erode at different rates. Soft rock is weathered and eroded faster than dense rock. The tall pillars of rock in Utah's Bryce Canyon are formed by the hard rock left behind after the soft rock has worn away.

The buildup of materials can also create landforms. Ocean waves pound rocks into small pieces of sand and then deposit the sand on the coastline. Rivers deposit soil where they empty into lakes and oceans.

Large masses of ice and snow can form valleys and plains. These masses are classified by size as glaciers, polar ice caps, or ice sheets. A **glacier** is the smallest ice mass; it moves slowly over land. Ice caps are high-altitude ice masses, and ice sheets extend more than 20,000 square miles over Greenland and Antarctica.

Human actions also change Earth. Coal mining has leveled mountains. Millions of acres of forests have been cut down and landslides and erosion have resulted. Canals have altered natural waterways. Pollution caused by burning gasoline and other fuels is released into the air and can lead to chemical erosion. Even buildings can be eroded by chemical pollutants.

Physical Geography

Lesson 2: A Changing Earth, *Continued*

Forces of Change	
Natural Forces	**Human Actions**
plate movements	cutting down forests
weathering	mining
erosion	digging canals
water and ice movement	pollution

Studies show that humans have changed the environment of Earth faster and more completely in the last 50 years than at any time in history. One major reason is demand for food and natural resources is greater than ever. That demand continues to grow as Earth's population increases.

Changes to Earth's surface caused by natural physical processes happen slowly, over millions of years. However, changes caused by humans can damage Earth's surface quickly. These changes can threaten our safety and survival. We need to take steps to protect the environment for future generations.

☑ **Reading Progress Check**

9. Categorize each of the following events as a slow change or a sudden change: earthquake, glacier, tsunami, volcano eruption, wind erosion, water erosion, plate movement.

_____

_____

_____

Writing

Check for Understanding

1. **Informative/Explanatory** What are two ways earthquakes contribute to changes on Earth's surface?

_____

_____

_____

_____

2. **Argument** Why should humans be careful about changing Earth's surface?

_____

_____

_____

_____

networks

Physical Geography

Lesson 3: Land and Water

ESSENTIAL QUESTION
How does geography influence the way people live?

Terms to Know
plateau a flat area that rises above the surrounding land
plain a large expanse of land that can be flat or have a gentle roll
isthmus a narrow strip of land that connects two larger land areas
continental shelf the part of a continent that extends into the ocean, then drops off sharply
trench a long, narrow, steep-sided cut on the ocean floor
desalinization a process that makes salt water safe to drink
groundwater the water contained inside Earth's crust
delta an area where sand, silt, clay, or gravel is dropped at the mouth of a river
water cycle the process in which water is used and reused on Earth
evaporation the change of liquid water to water vapor
condensation the change of water vapor to a liquid or solid state
acid rain rain that contains harmful amounts of poisons due to pollution

What Do You Know?
In the first column, answer the questions based on what you know before you study the lesson. After this lesson, complete the last column.

Now...		Later...
	Where is water vapor found on Earth?	
	Where is liquid water found on Earth?	
	Where is solid water found on Earth?	

Physical Geography

Lesson 3: Land and Water, *Continued*

Land Takes Different Forms

Guiding Question *What kinds of landforms cover Earth's surface?*

Earth has many different landforms. Scientists group landforms by characteristics. One characteristic they use is elevation. Elevation describes how far above sea level a landform is.

A **plateau** is a flat area that rises above the surrounding land. A **plain** is a large expanse of land that can be flat or have a gentle roll. A valley is a lowland area between two higher sides. Some valleys are small, while others are very large. Because water runs into them, valleys often have rich soil that is good for farming and raising livestock.

Landforms can also be described by their relation to water. Some landforms are surrounded by water. Continents are huge landmasses surrounded mostly by water. Islands are much smaller landmasses completely surrounded by water.

A peninsula is a long, narrow landform that extends into water. A peninsula is always connected to a larger landmass at one end. An **isthmus** is a narrow strip of land that connects two larger land areas. The Central American country of Panama is on an isthmus located at the narrowest point in the Americas.

The ocean floor also has different landforms. One type of ocean landform is the continental shelf. A **continental shelf** is an underwater plain that borders the edge of a continent. A continental shelf ends when it slopes downward to the ocean floor. Another type of ocean landform is a trench. A **trench** is a long, narrow, steep-sided cut on the ocean floor. The ocean floor also has mountains and volcanoes.

The Blue Planet

Guiding Question *What types of water are found on Earth's surface?*

Water can exist in different forms: solid, liquid, and gas. All three forms can be found all over the world.

Forms of Water	
solid	glaciers, polar ice caps, and ice sheets
liquid	rivers, lakes, and oceans
gas	water vapor in the atmosphere

Activating Prior Knowledge

1. On what type of landform do you live?

   _____

   _____

Marking the Text

2. Circle two types of landforms found on continents. Underline two types of landforms found on the ocean floor.

Activating Prior Knowledge

3. How is an island similar to a lake?

   _____

   _____

   _____

   _____

Reading Progress Check

4. How is a valley similar to an ocean trench?

   _____

   _____

   _____

   _____

   _____

   _____

Physical Geography

Lesson 3: Land and Water, *Continued*

Copyright © The McGraw-Hill Companies, Inc.

? Identifying

5. What is the body of salt water closest to where you live? What is the body of freshwater closest to you?

_____

_____

Inferring

6. Why is desalinization an important process?

_____

_____

_____

? Explaining

7. How is a delta formed?

_____

_____

_____

✓ Reading Progress Check

8. Describe three ways in which water affects your life.

_____

_____

_____

_____

_____

_____

Water at Earth's surface can either be freshwater or salt water. Salt water contains salt and other minerals. Most of Earth's water is salt water. Salt water makes up all of Earth's oceans and a few lakes and seas. Even though many plants and animals live in salt water, humans cannot drink salt water. **Desalinization** is a process that separates out most salt and minerals and makes salt water safe to drink.

Freshwater makes up only three percent of the water on Earth. Most freshwater is frozen in the Arctic and Antarctic. Liquid freshwater is found in lakes, rivers, ponds, swamps, and in the ground. Water contained inside Earth's crust is called **groundwater**. Groundwater is an important source of drinking water.

Earth's largest bodies of water are its four salt water oceans. From largest to smallest they are the Pacific, Atlantic, Indian, and Arctic. The oceans are all connected to one another.

The places where oceans meet landmasses have unique land features. A coastal area where ocean waters are partially surrounded by land is called a bay. Larger areas that are partially surrounded by land are called gulfs.

Freshwater rivers are found all around the world. Some rivers meet other rivers and join to make a larger river. A river's end point is called the mouth of the river. A **delta** is an area where sand, silt, clay, or gravel is dropped at the mouth of a river.

Bodies of water affect the people who live near them. People can fish for food. The ocean floor can be mined for minerals. Water is used for transportation, and also for sports and recreation.

The Water Cycle

Guiding Question *What is the water cycle?*

The amount of water on Earth has not changed since the planet was formed billions of years ago. Water is constantly recycled. It moves over, under, and above Earth's surface and changes form as it is recycled. This system is called the **water cycle.**

As the sun warms water on Earth, the liquid water changes into water vapor. This is called **evaporation**. Evaporation happens constantly around us. As water evaporates, tiny droplets of air rise into the atmosphere and gather into clouds. Eventually, the water vapor changes into the liquid or solid state. This is called **condensation**. Water then falls back to the Earth's surface as rain, snow, or hail. Liquid rainwater collects in rivers, lakes, oceans, and underground water supplies. In this way, water taken from Earth's surface during evaporation returns as precipitation.

Physical Geography

Lesson 3: Land and Water, *Continued*

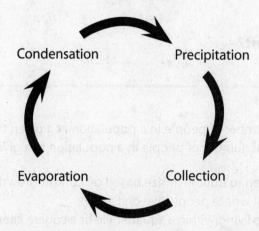

Condensation   Precipitation

Evaporation   Collection

Human actions have damaged the world's water supply. Toxic chemicals have polluted rivers, lakes, oceans, and groundwater. Some of the fuels we use to create energy release poisonous gases into the air. These gases combine with water vapor, which then falls to Earth as **acid rain**. Acid rain pollutes drinking water, erodes mountains and buildings, and kills plants and animals in lakes, rivers, and oceans.

Marking the Text

9. Circle the process by which water enters the atmosphere. Underline the process by which water vapor returns to Earth.

✔ Reading Progress Check

10. What are the causes and effects of acid rain?

_____

_____

_____

_____

Writing
Check for Understanding

1. **Narrative** Write a paragraph about the life of a child who lives on the coastline. Be sure to include details about how living near water affects the child's life.

_____

_____

_____

2. **Informative/Explanatory** If the amount of water on Earth has not changed, why should humans be worried about the world's water supplies?

_____

_____

_____

networks

Human Geography

Lesson 1: The World's People

ESSENTIAL QUESTION
How do people adapt to their environment?

Terms to Know

death rate number of deaths compared to total number of people in a population at a given time

birthrate number of babies born compared to total number of people in a population at a given time

doubling time number of years it takes a population to double in size based on current growth rate

population distribution the geographic pattern of where people live on Earth

population density the average number of people living within a square mile or a square kilometer

urban areas that are densely populated

rural areas that are lightly populated

emigrate to leave one's home to live in another place

immigrate to enter and live in a new country

refugees people who flee a country because of violence, war, or persecution

urbanization when cities grow larger and spread into nearby areas

megalopolis a huge city or cluster of cities with an extremely large population

---

Where in the World: India's Major Cities

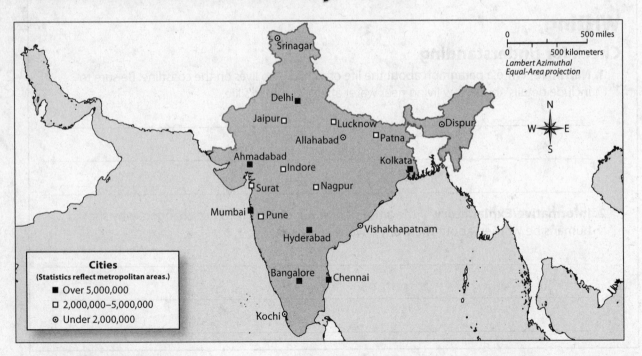

Human Geography

Lesson 1: The World's People, *continued*

Earth's Growing Population

Guiding Question *What factors contribute to Earth's constantly rising population?*

Since 1800, Earth's population has grown very large. Why has this happened? One reason is the falling **death rate**. The death rate is the number of deaths compared to the total number of people in a population at a given time. The death rate has gone down because of better health care, more food, and cleaner water.

Another reason is the rising **birthrate**. Birthrate is the number of babies born compared to the total number of people in a population at a given time.

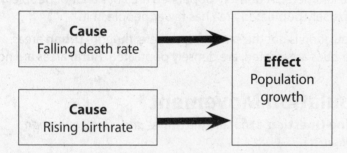

| Cause |
| Falling death rate |

| Cause |
| Rising birthrate |

Effect
Population growth

Over the past 60 years, the birthrate has been falling, even though the population as a whole has been growing. This is because the birthrate has been higher than the death rate. Fortunately, the overall growth rate has slowed down. The United Nations says that the world's population should reach its peak by 2050 before it begins to fall.

Populations grow at different rates in different parts of the world. This happens because families are influenced by their culture and religion. **Doubling time** is the number of years it takes a population to double in size based on its current growth rate.

A growing population can cause many problems for people. Cities and towns become too crowded. Water supplies become polluted, and diseases spread quickly. In addition jobs are hard to find, and families live in poverty.

A growing population can also cause problems for the environment. More fuel is needed for cars and homes, so miners drill for more energy resources. More farms are needed, so forests are cut down. More factories create air and water pollution.

Many groups are working hard to clean up the environment. One way is to use energy resources that do not cause pollution. These include solar, wind, and geothermal resources.

 Calculating

1. In 2000, a population of a city is one million people. If it has a doubling time of 25 years, when will its population reach two million?

_____

_____

Marking the Text

2. Read the text on the left. Then underline the problems that a growing population can have on its *people*. Circle the problems that a growing population has for its *environment*.

Reading Progress Check

3. What is the difference between birthrate and death rate?

_____

_____

_____

_____

_____

_____

Human Geography

Lesson 1: The World's People, *continued*

 **Defining**

4. What is the difference between population distribution and population density?

_____

_____

_____

 Reading Progress Check

5. Give one example of an urban area and one example of a rural area.

_____

_____

_____

? **Making Connections**

6. What is the difference between emigration and immigration?

_____

_____

_____

✏ **Marking the Text**

7. Circle the push factors that would cause people to move to a new area. Underline the pull factors that would cause people to move.

Population Patterns

Guiding Question *Why do more people live in some parts of the world than in others?*

How do people decide where to live? They might live somewhere to be close to work, for religious reasons, or because of transportation. Wherever they go, they want their basic needs to be met, including shelter, food, water, and jobs.

The geographic pattern of where people live is called **population distribution**. This is one way of looking at the pattern of a population. Another way is with density. **Population density** is the average number of people living within a square mile or square kilometer. A densely populated area has a lot of people in it. A sparsely populated area has fewer people in it.

Population is not the same everywhere, though. **Urban** areas, where cities are located, are densely populated. **Rural** areas are not.

Population Movement

Guiding Question *What are the causes and effects of human migration?*

As large numbers of people move into an area, the population increases. As people leave an area, the population decreases. Moving from one place to another is called migration. It is one of the main reasons that population changes in our world today.

There are two words to describe how people move: emigrate and immigrate. **Emigrate** is when people *leave* their home to live in another place. This can happen within the same nation, such as when people move from a village to a city. Often, though, it happens when people leave one nation and move to another. For instance, millions of people have emigrated from Europe to live in the United States. **Immigrate** is when people *enter* a new country to live there.

People leave one area and move to another area because of push-pull factors. The word *factor* means "reason," so a push factor is a reason that drives people from an area. It *pushes* them out.

A bad economy is a push factor. So are natural disasters, like floods or earthquakes. War is also a push factor. People who flee a country because of violence, war, or persecution are called **refugees**.

Pull factors attract people toward an area. They pull them in. A better job is a type of pull factor. So are friends, family, and a good university.

Human Geography

Lesson 1: The World's People, *continued*

When people move, it can affect many things in a region. The land, resources, culture and economy can all change.

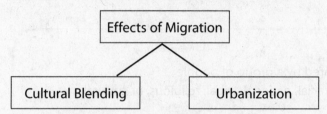

Effects of Migration
- Cultural Blending
- Urbanization

Cultural blending happens when people from different cultures move to the same place and live close together. Their cultures mix and blend. This can influence the art, music, clothing, food, and language of a place.

Urbanization happens when cities grow larger and spread into nearby areas. The reason for this is migration, particularly when people move into cities for jobs. As urbanization occurs, transportation and trade centers grow and more services are needed. This means jobs become available for people in the medicine, education, housing, and food sectors.

Urbanization is happening all over the world. In some areas, cities grow until they reach other cities and connect with them. A **megalopolis** is a huge city or cluster of cities with an extremely large population.

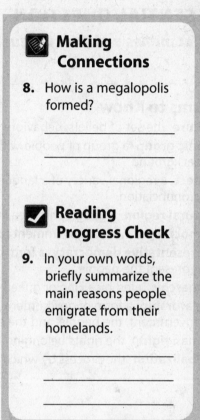

Making Connections

8. How is a megalopolis formed?

_____

_____

_____

Reading Progress Check

9. In your own words, briefly summarize the main reasons people emigrate from their homelands.

_____

_____

_____

Writing

Check for Understanding

1. Informative/Explanatory Is cultural blending a positive or negative effect of migration? Why?

_____

_____

_____

2. Narrative Imagine you have just emigrated from your homeland to another part of the world. Write a diary entry about why you moved. Describe some experiences in your new home.

_____

_____

_____

Human Geography

Lesson 2: Culture

ESSENTIAL QUESTION

What makes a culture unique?

Terms to Know

culture the set of beliefs, behaviors, and traits shared by a group of people

ethnic group a group of people with a common racial, national, tribal, religious, or cultural background

dialect a regional variety of a language with unique features, such as vocabulary, grammar, or pronunciation

cultural region a geographic area in which people have certain traits in common

democracy a type of government run by the people

representative democracy a form of democracy in which citizens elect government leaders to represent the people

monarchy ruled by a king or queen

dictatorship a form of government in which one person has absolute power to rule and control the government, the people, and the economy

human rights the rights belonging to all individuals

globalization the process by which nations, cultures, and economies become mixed

---

When did it happen?

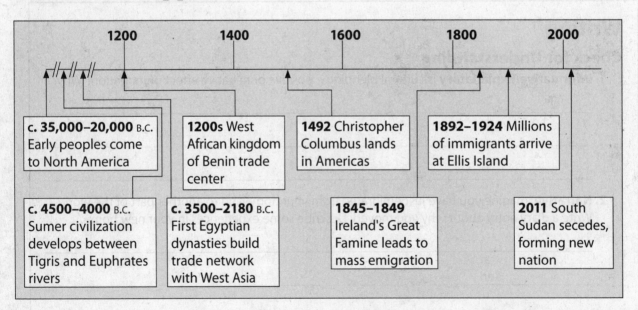

c. 35,000–20,000 B.C. Early peoples come to North America

c. 4500–4000 B.C. Sumer civilization develops between Tigris and Euphrates rivers

1200s West African kingdom of Benin trade center

c. 3500–2180 B.C. First Egyptian dynasties build trade network with West Asia

1492 Christopher Columbus lands in Americas

1845–1849 Ireland's Great Famine leads to mass emigration

1892–1924 Millions of immigrants arrive at Ellis Island

2011 South Sudan secedes, forming new nation

Human Geography

Lesson 2: Culture, *continued*

Elements of Culture

Guiding Question *How is culture part of your life?*

Culture is a set of beliefs, behaviors, and traits shared by a group of people. It is also the people of a certain culture. For instance, "the Hindu culture" could mean Hindu traditions. Or it could mean the people who follow those traditions. Or it could mean both.

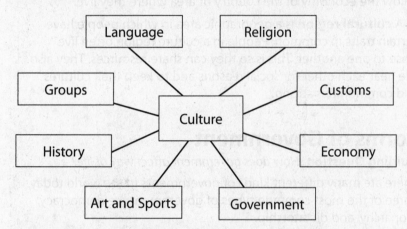

We can look at members of a culture in terms of age, gender, or ethnic group. An **ethnic group** is people who share a common racial, national, tribal, religious, or cultural background.

Language is a way of passing on cultural beliefs. Some languages are spoken all over the world. Other languages are spoken differently in different regions or by different ethnic groups. A **dialect** is a regional variety of a language. Dialects can have their own vocabulary, grammar, or pronunciation.

Religion influences how people of a culture see the world. Some people practice religion only on special occasions. Others see religion as a very important part of their life. Many cultures base their way of life on a religion.

Customs are an outward display of culture. Some are used formally. Others show good manners or are a sign of respect. When people greet each other in the United States they shake hands. In Japan people bow to each other. In France and Italy they will greet each other with a kiss on the cheek. These are all customs of the individual cultures.

The arts are an important part of a culture. This includes dance, music, visual arts, and literature. Music and dance can be part of religious ceremonies or cultural events. Art can also be a form of personal expression. Sports are also a part of culture. Many sports we play today began with sports played by other cultures.

Describing

1. What are the two ways that culture can be defined?

_____

_____

_____

_____

Defining

2. How is dialect different from language?

_____

_____

_____

_____

Explaining

3. How is religion a part of culture?

_____

_____

_____

Making Connections

4. Describe one way that you could participate in the arts in your community.

_____

_____

_____

_____

_____

Marking the Text

5. Highlight the definition of a cultural region.

Reading Progress Check

6. What cultural traditions do you practice? Make a list of the beliefs, behaviors, languages, foods, art, music, clothing, and other elements of culture that are part of your daily life.

_____

_____

_____

Marking the Text

7. Read the text on the right and underline the reasons why dictators can abuse power.

Reading Progress Check

8. In your own words, describe the system of government used in the United States.

_____

_____

_____

Government is also an element of culture. Governments maintain order within a country and protect it from other countries. They provide services to citizens, such as education. Different countries have different systems of government.

Economies control the use of natural resources. They also define how goods are made and distributed to people. Most cultures follow the economy of the country or area where they live.

A **cultural region** is a geographic area in which people have certain traits in common. People in a culture region often live close to one another. This is so they can share resources. They also live near each other for social reasons and to keep their cultures and communities strong.

Forms of Government

Guiding Question *How does government affect way of life?*

There are many different kinds of governments in the world today. Three of the most common types of government are democracy, monarchy, and dictatorship.

A **democracy** is a type of government run by the people. Therefore, people hold the power. Their rights and freedoms are protected. In addition, they can propose and vote on laws and policies for their nation.

In some democracies, the people elect government leaders to represent them. These leaders then make and carry out laws and policies for the people. This is called a **representative democracy**. The United States is a representative democracy.

A **monarchy** is ruled by a king or queen, called a monarch. In this type of government, power and leadership are passed down within a family. Older generations pass power down to younger generations. In the past, monarchs had complete power. Now, they often stand for a country's traditions and values. Elected leaders do the work of running the government.

A **dictatorship** is a form of government in which one person has absolute power. The dictator rules and controls everything: the government, the people, and the economy. People have few rights. The dictator can make laws without worrying about how fair or practical they are.

Some dictators abuse their power. This can limit personal freedoms and human rights. Human rights are the rights belonging to all individuals. These include life, liberty, security, privacy, and the right to marry and have children.

Human Geography

Lesson 2: Culture, *continued*

Changing Cultures

Guiding Question *How do cultures change over time?*

Cultures change over time. As people move from place to place, they bring their cultural traditions with them. These traditions blend with the cultures of the places where they move.

Trade can also change a culture. When people trade, they travel to other places. Once there, they sell and exchange goods, bringing their languages, customs, and ideas with them. In turn, they take these same things back home.

Culture can also change because of technology. The telephone and email have made it easier to communicate. Television and the Internet have made it easier to find information and learn about new ideas.

Today's world is becoming more blended every day. As this happens, new cultural traditions are born. **Globalization** is when nations, cultures, and economies become mixed. This can have both positive and negative effects on cultures.

Positive Effects of Globalization	Negative Effects of Globalization
Creates more understanding and acceptance of other cultures	Produces tension and conflict
Helps spread ideas and innovations	Causes some cultures to be damaged or destroyed

Marking the Text

9. Find three ways that cultures change over time and highlight them with a marker.

 Interpreting

10. Think of one way technology has helped you learn about a new culture and describe it.

_____

_____

_____

☑ Reading Progress Check

11. What is globalization?

_____

_____

Writing

Check for Understanding

1. **Narrative** If you were to trade something you own for something from another country, what would you choose? How might this affect the culture of the other country?

_____

_____

2. **Narrative** Give examples from your own experience of how cultures can change over time.

_____

_____

network⊛rks

Human Geography

Lesson 3: The World's Economies

ESSENTIAL QUESTION

Why do people make economic choices?

Terms to Know

renewable resource a resource that can be totally replaced or is always available naturally

nonrenewable resource a resource that cannot be totally replaced

economic system how a society decides on the ownership and distribution of its economic resources

traditional economy an economy in which resources are distributed mainly through families

market economy an economy in which most of the means of production are privately owned

command economy an economy in which the means of production are publicly owned

mixed economy an economy in which parts may be privately owned, and parts may be owned by the government

gross domestic product (GDP) the total dollar value of all final goods and services produced in a country during a single year

standard of living the level at which a person, group, or nation lives as measured by the extent to which it meets its needs

productivity measurement of what is produced and what is required to produce it

export when a country sends a product to another country

import when a country brings in a product from another country

free trade when a group of countries decides to set few or no tariffs or quotas

sustainability the economic principle by which a country works to create conditions where all the natural resources for meeting the needs of society are available

What Do You Know?

List a type of economic system in each outer circle. Then fill in what you learn about them from the lesson.

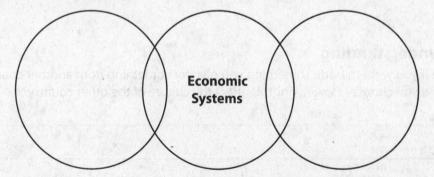

Economic Systems

The Basic Economic Question

Guiding Question *How do people get the things they want and need?*

People meet their wants and needs by using resources.

Human Geography

Lesson 3: The World's Economies, *continued*

Resources that come from the earth are called natural resources. These include water, soil, plants, and animals. Resources can also come from humans. These include work, skills, and talents.

One need that everyone shares is energy. Energy is the power to do work. It comes from two basic types of energy resources: **renewable** and **nonrenewable**.

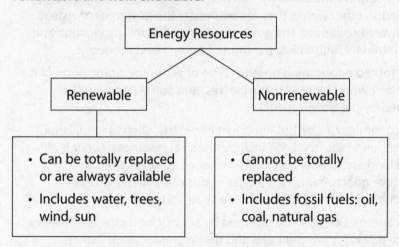

Energy Resources

Renewable

Nonrenewable

- Can be totally replaced or are always available
- Includes water, trees, wind, sun

- Cannot be totally replaced
- Includes fossil fuels: oil, coal, natural gas

Energy is not just a need, however. It is also sometimes used to power things that are not needed but wanted. However, there are not always enough resources to fill both the needs and the wants. Therefore, we must make choices. We must decide whether to use renewable or nonrenewable resources.

One way of making this decision is to weigh the opportunity cost of both resources. The opportunity cost is the value of what we must give up to get something else.

Countries Build Economies

Guiding Question *What kinds of economic systems are used in our world today?*

Societies make, or produce, goods and services. They also distribute them to people and decide how these goods and services are to be used. This describes the economic resources of a society.

How a society takes ownership and distribution of its economic resources is called its **economic system**. There several different types of economic systems.

One type is a **traditional economy**. In this economy, resources are distributed mainly through families. Farming and herding are often traditional economies.

? **Drawing Conclusions**

1. How can energy be both a want and a need?

   _____

   _____

A♭c **Defining**

2. Underline the definition of opportunity cost. Then provide an example of this from your own life.

   _____

   _____

✓ **Reading Progress Check**

3. How are renewable and nonrenewable resources alike, and how are they different?

   _____

   _____

✎ **Identifying**

4. In which type of economic system are the means of production owned by families?

   _____

   _____

Lesson 3: The World's Economies, *continued*

Copyright © The McGraw-Hill Companies, Inc.

Marking the Text

5. The primary sector involves which part of the economy: land, labor, capital, or industry. Circle your answer.

Describing

6. In your own words, describe what gross domestic product is.

_____

_____

_____

Reading Progress Check

7. How is standard of living a sign of economic performance?

_____

_____

_____

Defining

8. What is the difference between an export and an import?

_____

_____

_____

_____

In a **market economy**, private businesses own most of the means of production. These businesses decide how much to produce and distribute. To do that, they look at sales. They also look at the demand for products and resources. Capitalism is another name for a market economy.

In a **command economy** the means of production are publicly owned. In other words, they are owned by the government instead of private businesses. The government also controls production and distribution. Communism is a form of a command economy.

A **mixed economy** is another type of economy. Some parts of it may be owned by private businesses, and some parts may be owned by the government.

Economies can be broken down into parts, such as land, labor, capital, and industry. *Land* includes natural resources. *Labor* is all paid workers. *Capital* is all human-made resources used to produce goods. *Industry* is a type of business. Banking, retail, agriculture, and transportation are types of industries.

Economies can also be grouped by activity, or sector. The *primary sector* produces raw materials and the *secondary sector* makes finished goods. The *tertiary (third) sector* is the service industry.

How well is an economy doing? To find out, we look at economic performance. One way of seeing economic performance is with **gross domestic product (GDP)**. This is the total dollar value of all final goods and services made in a country during a single year.

The **standard of living** is another way of looking at performance. This is the level at which a person, group, or nations lives. It is measured by how well a person meets his or her needs. **Productivity** is a measurement of what is produced and what is required to produce it.

National economies can be classified by types. Developed countries are industrialized. Developing countries are less industrialized. They often have weak economies. In between are newly industrialized countries. They are in the process of becoming industrialized.

Global Economy

Guiding Question *How do the world's economies interact and affect one another?*

Trade is the business of buying, selling, or bartering. Countries trade when they send, or **export**, a product they make to another country. A country can also bring in, or **import**, a product.

Human Geography

Lesson 3: The World's Economies, *continued*

Sometimes it costs extra to import a product. This extra cost is called a tariff. Governments add tariffs so people will buy products made in their own country. Governments might also set limits on how many goods can be imported. However, a group of countries may decide to set few or no tariffs or limits. This is called **free trade**.

Trade can help build economic growth. It can increase a nation's income. On the other hand, it can cost jobs. When a country imports something, it doesn't need to make it. There are many economic organizations concerned with trade.

World Trade Organization (WTO)	Helps regulate trade between nations
World Bank	Provides financing, advice, and research to developing nations
International Monetary Fund (IMF)	Monitors economic development and lends money to nations in need
European Union (EU)	Group of European countries that have one economic unit and currency

Nations can produce their own goods, or they can trade for them. However, it is important that they have all the natural resources they need to meet the needs of society. When a country can do this, it is practicing **sustainability**. Sustain means maintain. The country is maintaining the resources it needs.

Explaining

9. Why would a government want to limit the number of imports coming into a country?

_____

_____

Reading Progress Check

10. How might globalization affect developing countries?

_____

_____

_____

Writing

Check for Understanding

1. Informative/Explanatory Why do you think governments might want free trade?

_____

_____

2. Argument Write a paragraph explaining the benefits of one economic system mentioned in the lesson. Make a strong argument as to why this system is better than the others.

_____

_____

_____

_____

networks

The United States East of the Mississippi River

Lesson 1: Physical Features

ESSENTIAL QUESTION
How does geography influence the way people live?

Terms to Know
subregion smaller part of a region
lock gated passageway used to raise or lower boats in a waterway
tributary small river that flows into a larger river
levee raised riverbank used to control flooding
coastal plain flat lowland along a coast
fall line area where waterfalls flow from higher to lower ground
hurricane storm with strong winds and heavy rains

Where in the World: The United States East of the Mississippi River

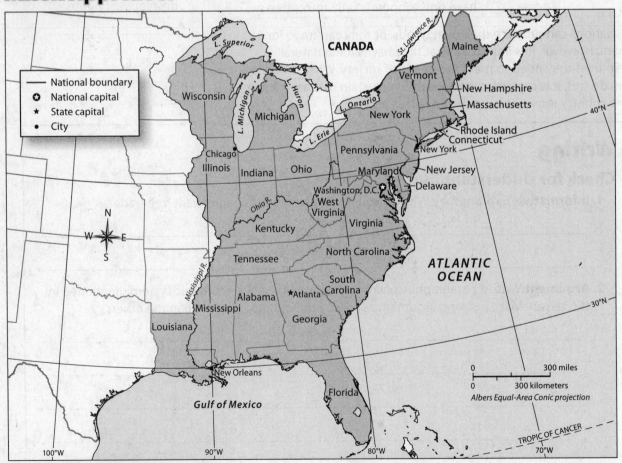

The United States East of the Mississippi River

Lesson 1: Physical Features, *Continued*

The Regions

Guiding Question *How do the physical features of the United States east of the Mississippi River make the region unique?*

The United States is very large. There are many differences between its various parts. It is made up of geographic areas geographers call regions. Each region has features that make it different from other regions. The regions can be divided into smaller parts, called **subregions**.

We can divide the United States into two regions. One region is east of the Mississippi River, and the other is west of the Mississippi. In this lesson, you will learn about the United States east of the Mississippi. It is made up of four subregions. They are the New England, the Mid-Atlantic, the Midwest, and the Southeast subregions.

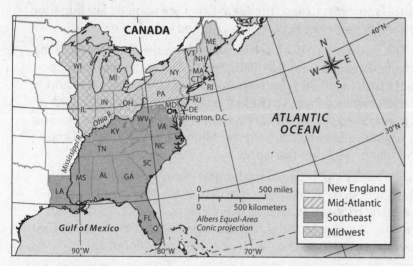

Many of the first English colonists settled in New England. They arrived during the 1600s. The settlers named the area to honor their homeland.

The Mid-Atlantic states were part of the original thirteen colonies. Our nation's capital, Washington, D.C., is located in the Mid-Atlantic subregion.

The Midwest states share borders with one or more of the Great Lakes. A large part of America's food crops are grown in the Midwest's rich soil.

The Southeast is the largest subregion in the United States east of the Mississippi. Some of these states have long coastal borders on the Atlantic Ocean or Gulf of Mexico.

Defining

1. What is the definition of a *subregion*?

   _____

   _____

Marking the Text

2. Read the text on the left. Highlight the names of the subregions of the United States east of the Mississippi each time you see them.

Explaining

3. How is each of the four subregions divided into smaller areas?

   _____

   _____

   _____

Reading Progress Check

4. Why do you think geographers divide the United States at the Mississippi River instead of dividing it through the middle into equal halves?

   _____

   _____

networks

The United States East of the Mississippi River

Lesson 1: Physical Features, *Continued*

Marking the Text

5. Read the text on the right. Highlight the names of all the bodies of water that border the United States east of the Mississippi.

Describing

6. Why is the St. Lawrence Seaway important?

_____

_____

_____

_____

_____

✓ Reading Progress Check

7. How could a logging company in Kentucky send logs to a buyer on the Gulf of Mexico using an all-water route?

_____

_____

_____

_____

_____

_____

_____

Bodies of Water

Guiding Question *Which of North America's major bodies of water are located east of the Mississippi?*

The United States east of the Mississippi River is nearly surrounded by water. The largest body of water is the Atlantic Ocean. It borders the states along the East Coast. The East Coast stretches more than 2,000 miles (3,219 km), from Maine to Florida.

The Gulf of Mexico extends from Florida to Texas and Mexico. The land around the Gulf is called the Gulf Coast. Currents flow through the Gulf of Mexico like underwater rivers. One of these is the Gulf Stream. It also flows through the Atlantic Ocean.

The Great Lakes are a cluster of five lakes located in the Midwest and central Canada. They were formed by glaciers thousands of years ago. They hold more freshwater than any other location on Earth. From west to east, the Great Lakes are named Lake Superior, Lake Michigan, Lake Huron, Lake Erie, and Lake Ontario.

The St. Lawrence River carries water from the Great Lakes to the Atlantic Ocean. During the 1950s, the United States and Canada built canals, artificial waterways, to connect the Great Lakes and the St. Lawrence River. **Locks** in the canals allow water levels to rise and fall so that boats can move from one level to another. The St. Lawrence Seaway connects the Midwest to the Atlantic Ocean and seaports all over the world.

The Mississippi is one of the longest rivers in North America. It runs south from Minnesota to the Gulf of Mexico. The Mississippi has many **tributaries,** smaller rivers that flow into it. These rivers are busy waterways. In the past, the Mississippi often flooded. The government built walls called **levees** to control flooding.

The Ohio River is the Mississippi's largest tributary. It begins in western Pennsylvania. Ohio, Indiana, and Illinois are north of the river. West Virginia and Kentucky are on its south side. Like the Mississippi, it is an important shipping and transportation route. It connects much of the Midwest to the Mississippi.

Physical Landscape

Guiding Question *What characteristics make the physical landscape east of the Mississippi unique?*

Along the Atlantic Ocean and Gulf of Mexico, a **coastal plain** stretches from the northeast U.S. to Mexico. This is a flat lowland area. Parts of it are below sea level and they often become flooded by storms and heavy rains.

The United States East of the Mississippi River

Lesson 1: Physical Features, *Continued*

The Appalachian Mountain system stretches from Alabama northeast to Canada. Dense forests cover much of the Appalachian Mountains. The mountains here stand side by side in parallel ranges. These ranges include the Blue Ridge Mountains in Virginia and the Great Smoky Mountains in Tennessee.

A **fall line** runs from New Jersey to South Carolina. It is a long, low cliff running parallel to the Atlantic coast. There are many waterfalls along this line. The fall line is a boundary between upland areas and the coastal plain.

The climate of the eastern United States is varied. New England and the Midwest see dramatic seasonal changes. They have cold winters and hot, humid summers. Coastal areas have milder climates. Much of the Southeast has a humid subtropical climate. Summers are rainy and hot. Winters are cooler and drier. **Hurricanes,** strong ocean storms, can strike along the coast.

The region has many minerals and energy resources. Minerals include iron ore, which can be made into metals. Metals are used in manufacturing and construction. Energy resources include coal, oil, and natural gas. Burning coal can produce electricity. There is a huge demand for mineral and energy resources. Mining them is a major industry.

The rich soil of the region is a valuable resource. It is excellent for growing crops such as grains, fruits, and vegetables. Industries such as logging and fishing are also important. Factories in the region produce automobiles, electronics, and clothing.

Defining

8. What is a fall line?

_____

_____

Marking the Text

9. Read the text on the left. Highlight the names of resources found in the United States east of the Mississippi.

Reading Progress Check

10. Why is farmland considered a natural resource?

_____

_____

Writing

Check for Understanding

1. **Informative/Explanatory** Choose one of the subregions of the United States east of the Mississippi River. Using information from this lesson, describe the region as completely as possible.

_____

_____

2. **Informative/Explanatory** Give an example of the effect waterways have on life in the United States east of the Mississippi.

_____

_____

The United States East of the Mississippi River

Lesson 2: History of the Region

ESSENTIAL QUESTION
Why is history important?

Terms to Know
indigenous living or existing naturally in a particular place
colonists people sent to live in a new place and claim land for their home country
agriculture growing crops and raising livestock
industry manufacturing and making products to sell

When did it happen?

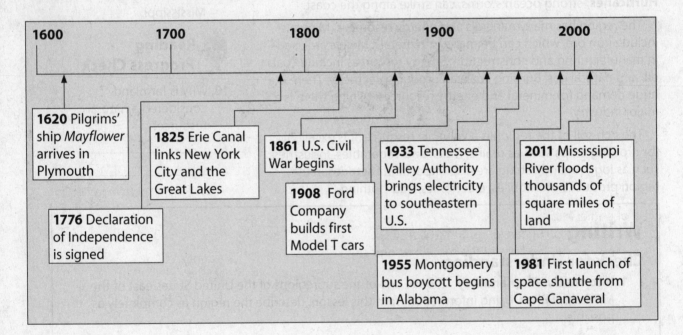

| 1600 | 1700 | 1800 | 1900 | 2000 |

1620 Pilgrims' ship *Mayflower* arrives in Plymouth

1825 Erie Canal links New York City and the Great Lakes

1861 U.S. Civil War begins

1933 Tennessee Valley Authority brings electricity to southeastern U.S.

2011 Mississippi River floods thousands of square miles of land

1776 Declaration of Independence is signed

1908 Ford Company builds first Model T cars

1955 Montgomery bus boycott begins in Alabama

1981 First launch of space shuttle from Cape Canaveral

networks

The United States East of the Mississippi River

Lesson 2: History of the Region, *Continued*

Early America

Guiding Question *Who were the first peoples to live in the United States east of the Mississippi?*

Native American peoples were the first humans to settle in North America. Each group had its own language, religion, and lifestyle. The Cherokee, the Iroquois, the Miami, and the Shawnee settled east of the Mississippi River. They are considered **indigenous** to the region, meaning that they were descended from the first people to live there.

Native Americans satisfied their needs by using the plants, animals, stones, water, and soil around them. Their ways of life were shaped by the environment. Peoples who lived in wooded areas made homes of bark and wood. They burned wood for heat. They hunted for food and used animal skins for clothing.

Native peoples built shelters suited to the climates where they lived. People who lived in hot climates built open-air homes. Other groups built solid structures and mound cities.

For most of their history, Native Americans lived isolated from people of other parts of the world. They lived off the land for thousands of years. They had little impact on the natural environment. Suddenly, the land they relied on was taken from them. When the first Europeans arrived, native peoples' ways of life changed forever.

People across the Atlantic Ocean, in Europe, grew interested in the Americas. They heard tales of these wild, rich lands. Explorers told of endless forests, rivers full of fish, and mountains filled with gold and silver. European kings and queens wanted to claim land in North America and control the natural resources.

Early European Settlements in What Is Now the United States		
Location	St. Augustine, Florida	Jamestown, Virginia	Plymouth, Massachusetts
Settled by	Spain	England	England
Date	1565	1607	1620

The first English **colonists** settled along the Atlantic Coast. Over the years, more Europeans came to America. By 1760, an estimated 1.7 million colonists lived in America. Large cities such as Boston and New York City started as small settlements. By 1750, there were thirteen English colonies in North America.

Marking the Text

1. Read the text on the left. Highlight the names of Native American peoples who lived east of the Mississippi River.

Defining

2. What does the word *indigenous* mean?

_____

_____

_____

Identifying

3. Where did the first Europeans who settled in what is now the United States come from?

_____

_____

_____

Summarizing

4. How did the arrival of European settlers change the life of the indigenous peoples east of the Mississippi River?

_____

_____

_____

_____

_____

The United States East of the Mississippi River

Lesson 2: History of the Region, *Continued*

? Explaining

5. Why did the American colonists declare independence from Great Britain?

_____

_____

_____

✓ Reading Progress Check

6. Why did European nations want to control land in North America?

_____

_____

_____

? Describing

7. What was the Land Ordinance of 1785?

_____

_____

_____

✓ Defining

8. What makes the United States east of the Mississippi River a good region for *agriculture*?

_____

_____

_____

_____

These colonies were controlled by English rulers thousands of miles away. The colonists did not like the laws and taxes the British government forced on them. In 1776 American colonists declared their independence and fought the Revolutionary War. The war ended in 1781, when the British surrendered. The 13 colonies were free. They called their new country the United States of America.

Settling the Land

Guiding Question *How has the movement of people shaped the culture of the United States east of the Mississippi?*

The new nation stretched from the Atlantic Ocean to the Mississippi River. However, little was known about the land west of the Appalachian Mountains. As more settlers arrived, people began to move inland. They wanted to claim land and build new lives.

Between 1700 and 1800, thousands of people settled along the Mississippi River. The Mississippi formed a natural boundary. The U.S. government wanted to claim as much of America's land as possible. Laws such as the Land Ordinance of 1785 governed settlement in new lands. Native Americans living on lands in the Ohio Country were forced to leave.

Land Ordinance of 1785

- Gave the United States legal claim to the Ohio Country (land north of the Ohio River and east of the Mississippi River).
- Settlers could buy sections of land for $1 per acre.
- Set aside land to be used for schools.

Farms in this region produced much of our country's fruits, vegetables, grains, and cotton. Good soil and frequent rain made the area good for **agriculture**. Planting and harvesting crops was hard work, done mainly by hand. Over the years, machines began to replace human workers. Farmers could plant and harvest more crops. However, fewer people were needed to work on farms. Thousands of people moved to cities looking for work.

New technology led to jobs in factories for millions of people. This is known as **industry**. Industry is an important part of the economy of this region. Thousands of factories have been built. They produce many goods, including cars. Some factories process foods and bottled drinks. This has been one of the world's leading industrial regions for over two hundred years.

The United States East of the Mississippi River

Lesson 2: History of the Region, *Continued*

People came to America to be free. They sought jobs, education, and other opportunities. Immigration has led to an amazing variety of languages, religions, cultures, and customs east of the Mississippi. These traditions have made America a unique and diverse nation.

In the 1830s, gold was discovered in the Southeast. Settlers poured in. Most did not find gold, but many stayed to start cotton farms. U.S. citizens wanted these lands. The government forced the Cherokee to leave these lands, which had been their home.

In 1838, the Cherokee were forced to migrate to Indian Territory in Oklahoma. This was far west of the Mississippi River. Thousands of Cherokee died as a result of the journey. This event became known as the Trail of Tears. This was one of many forced migrations in America's history. As America gained territory, Native Americans lost their lands and ways of life.

After the Civil War, slavery became illegal in all states. Yet many states passed laws to take away the new rights of the freed Americans. During the late 1800s, thousands of African Americans moved to states in the Mid-Atlantic, New England, and the Midwest. This relocation of people from the South to the North is called the Great Migration.

Rural to urban migration increased during the 1900s. Millions of people moved to cities to work in factories. It was one of the largest migrations in America's history.

✏ Marking the Text

9. Read the text on the left. Highlight the ways in which people have made America a diverse nation.

☑ Reading Progress Check

10. How did the invention of farm machinery lead to unemployment in the eastern United States?

_____

_____

_____

_____

Writing
Check for Understanding

1. List three ways in which Native Americans lives were shaped by their environment.

_____

_____

_____

2. What similarities are there between the Land Ordinance of 1785 and the Trail of Tears?

_____

_____

_____

netw⊕rks

The United States East of the Mississippi River

Lesson 3: Life in the Region

ESSENTIAL QUESTION

What makes a culture unique?

Terms to Know

metropolitan having to do with a large city

tourism industry that provides services to people who are traveling for enjoyment

civil rights basic rights that belong to all citizens

Rust Belt area of the Midwest, Mid-Atlantic, and New England where many factories closed during the 1980s

service industry businesses that provide services rather than products

What Do You Know?

In the first column, answer the questions based on what you know before you study. After this lesson, complete the last column.

Now...		Later...
	Why do people choose to live in metropolitan areas?	
	Why does the United States have three branches of government?	
	How do government actions affect people?	
	What are service industries?	

[Abc] Defining

1. What are the names of the major *metropolitan* areas located in your state?

   _____

   _____

Major Metropolitan Areas

Guiding Question *What is it like to live in a large metropolitan area in the United States east of the Mississippi River?*

People in the United States live in many different environments. Farmers and ranchers live in rural areas. Many people live in suburbs, where they can own pieces of land. Others prefer **metropolitan** areas, or large cities, such as New York, Boston, and Miami. Such cities are centers of culture, education, business, and recreation. The **tourism** business in cities provides jobs in service industries such as restaurants, hotels, and travel agencies. The money tourists spend brings revenue to state and local economies.

The United States East of the Mississippi River

Lesson 3: Life in the Region, *Continued*

New York and Chicago are two of the largest metropolitan areas east of the Mississippi River. New York City is on the Atlantic coast. It has the largest population of any city in the United States. It began as a Dutch colonial port. It still has one of the busiest ports in North America. New York City has five areas called boroughs. The smallest is Manhattan. It is the cultural, political, and economic center of the city. It is home to people of every kind of background. It is also a center of finance, advertising, entertainment, and the arts. More than 30 million tourists visit New York each year.

Chicago began as a small settlement on Lake Michigan. By the mid-1800s, it was the center of all railroad travel in the United States. Its importance as a transportation center grew with the opening of the St. Lawrence Seaway in the mid-1900s. It is still one of the nation's most important centers of shipping, transportation, and industry.

Atlanta is the capital of Georgia. With the coming of railroads, it became an economic, cultural, and political center of the South. It also became a transportation hub. It was burned to the ground during the Civil War, but rebuilt after the war. Today, it is the commercial center of the modern South. Many industries have headquarters in Atlanta.

New Orleans began as a port along the Mississippi River near the Gulf of Mexico. It was an important center of transportation and trade. It has also served as a major stop along supply lines during times of war. In 2005, the city was badly flooded by Hurricane Katrina. New Orleans is famous for its rich cultural traditions, including spicy foods, musical styles, and celebrations such as Mardi Gras. It was one of the first centers of jazz music. Tourism is very important to the economy of New Orleans.

The U.S. Government

Guiding Question *How have the government's actions affected the land and people of the United States?*

After it declared independence from Great Britain, the United States needed a system of government. Americans wanted a government to protect the rights and freedoms of the people. The government they created was a democracy. In a democracy, people elect representatives to run the government.

Representatives from each of the 13 states met in 1787 to write a plan of government. Their plan, the Constitution, is still the law of our country. The United States is a federal republic. This means that the national government shares power with the states.

✏️ Marking the Text

2. Read the text on the right. Highlight the names of major cities of the United States east of the Mississippi.

❓ Explaining

3. Why is tourism an important industry in many cities?

_____

_____

_____

☑️ Reading Progress Check

4. How are New York and New Orleans alike? How are they different?

_____

_____

_____

_____

✏️ Marking the Text

5. Read the text on the left. Underline the definition of a democracy. Then circle the definition of a federal republic.

The United States East of the Mississippi River

Lesson 3: Life in the Region, *Continued*

? Identifying

6. What is the Bill of Rights?

_____

_____

_____

_____

_____

_____

✓ Reading Progress Check

7. Why is the U.S. government divided into three branches?

_____

_____

_____

_____

_____

A♭c Defining

8. What is the definition of the term *civil rights*?

_____

_____

_____

_____

_____

_____

Amendments, or changes to the Constitution, have been made to meet changing needs. The first 10 amendments are known as the Bill of Rights. They guarantee the basic rights of citizens.

The men who wrote the Constitution did not want a single person or group to have all the power. They split the government into three branches: legislative, executive, and judicial. They are separate but equal. Each must work with the other two branches to govern the country. This is called a balance of power.

The Three Branches of Government		
Legislative	**Executive**	**Judicial**
Congress	President	Supreme Court
Makes laws for the nation	Carries out the nation's laws	Decides if laws are fair

Congress has two parts, the House of Representatives and the Senate. Members are elected by the people and come from all 50 states. Both must agree on a law before it is passed. Laws have been made to protect the land. However, some government actions have harmed the land by allowing companies to dig for coal and drill for oil. This has polluted rivers and leveled mountains.

Government actions have sometimes harmed people as well. **Civil rights**, the basic human rights that belong to all citizens, have not always been respected for Native American and African American people. To answer African Americans' demand for their rights, the federal government made laws ending segregation, or the enforced separation of people by race. While most governments obeyed these laws, some local governments refused.

Other government actions, such as building roads and bridges and providing aid for people in need, help people and enrich our lives. We should remember that no government is perfect. The U.S. government is made up of many different people doing many different jobs.

Everyday Life

Guiding Question *How has diversity shaped the culture of the United States?*

The religion with the largest number of followers in the United States is Christianity. However, there are also people who practice the Jewish, Muslim, and other faiths. Sixteen percent of Americans do not claim a religion.

networks

The United States East of the Mississippi River

Lesson 3: Life in the Region, *Continued*

The U.S. Census classifies the American population into several categories shown in the chart below.

Population of the United States, 2010			
Caucasian	72%	Native American	1%
African American	12%	Two or more races	3%
Asian American	5%	Pacific Islanders or others	7%

The United States has one of the largest and strongest economies in the world. Agriculture has long been an important industry. Today, family farms are being replaced by corporate farms.

In the late 1800s and early 1900s, the U.S. moved from an agricultural economy to an industrial economy. Factories producing cars, appliances, machinery, and other items have employed millions of people. In the 1980s, the manufacturing economy began to weaken. Many people lost their jobs. So many factories closed in the Midwest, Mid-Atlantic, and New England that those areas were called the "**Rust Belt**."

In recent years, more Americans have found jobs in **service industries**. They provide services rather than products. Restaurants, electricians, and auto repair shops are service industries. Industries such as finance, insurance, and education also provide jobs and revenue. Different types of industries are important to the economy of the eastern United States. Today the economy is shifting to be more connected to the computer-information age.

Making Connections

9. Read the text on the right. Highlight the names of service industries.

Reading Progress Check

10. What is the main difference between service industries and manufacturing industries?

_____

_____

_____

_____

_____

Writing

Check for Understanding

1. What do all of the metropolitan areas mentioned in this chapter have in common?

_____

_____

2. What types of businesses and industries are important to the economy of the United States east of the Mississippi today?

_____

_____

_____

networks

The United States West of the Mississippi River

Lesson 1: Physical Features

ESSENTIAL QUESTION
How does geography influence the way people live?

Terms to Know

cordillera a region of parallel mountain chains
timberline the elevation above which it is too cold for trees to grow
contiguous joined together inside a common boundary
Continental Divide an imaginary line through the Rocky Mountains that separates rivers that flow west from rivers that flow east
irrigation the process of collecting water and using it to water crops
chinook a dry wind that sometimes blows over the Great Plains in winter
ethanol a liquid fuel made in part from plants
national park a park that has been set aside for the public to enjoy for its great natural beauty

Where in the World: U.S. West of the Mississippi River

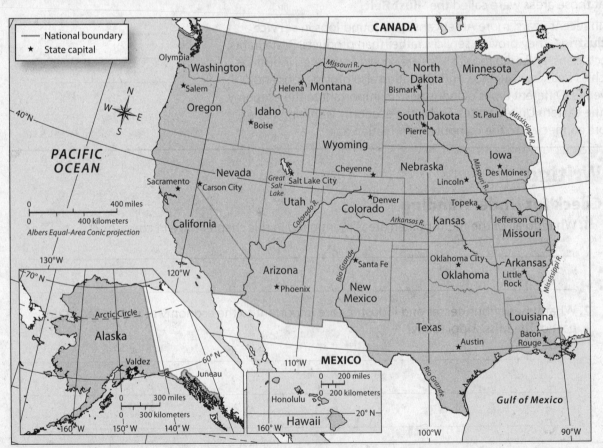

The United States West of the Mississippi River

Lesson 1: Physical Features, *continued*

Physical Landscape

Guiding Question *How do the physical features of the United States west of the Mississippi River make the region unique?*

The Mississippi River divides the United States into two parts. The area west of the Mississippi is larger than the area to the east. States west of the Mississippi are usually larger than states east of the river. Their populations are usually more spread out than in the east. The landforms are steeper and rockier. The climate tends to be drier. Many are rich in natural resources.

The Great Plains are just west of the Mississippi. The Plains appear flat in many places. In other places they look like gently rolling land. The Great Plains were once covered by wild grasses. Herds of bison grazed there. Today the Great Plains are covered by farms and ranches. The Black Hills are in the northern Great Plains. Evergreen trees darken the hills and give them their name.

The Rocky Mountains are located on the western edge of the Great Plains. The Rockies are not a single mountain chain but a **cordillera**. A cordillera is a region of parallel mountain chains. The Rockies extend from Canada to Mexico. Some peaks are as high as 14,000 feet. Trees cover the slopes up to the **timberline**, the elevation where it is too cold for trees to grow.

Several different mountain ranges line the Pacific Coast. Many of these mountains formed where tectonic plates meet. The lithosphere cracked. Broken land rose into steep rugged mountains. Heavy rainfall and cold temperatures combine to create glaciers in the mountains.

The Cascades run through western Oregon and Washington State. The Cascades were made from volcanic activity, and some of these volcanoes are still active today.

The Sierra Nevada range runs along the California-Nevada border. Mount Whitney is in this mountain range. Mount Whitney rises to about 14,500 feet (4,419 m). It is the highest point in the 48 contiguous states. **Contiguous** means "connected to." The contiguous states are those from the Atlantic to the Pacific Oceans. Hawaii and Alaska are not among the contiguous states.

Between the Rockies and the Cascade Range and Sierra Nevada is a variety of landforms. In the southeast is the Colorado Plateau. A plateau is a large area of generally flat land that is at a high elevation. Many canyons cut through the Colorado Plateau. One of these canyons is the Grand Canyon. The Colorado River winds along the bottom of the canyon, more than a mile below the rim.

? **Identifying**

1. Which state is likely to be larger, Massachusetts or Idaho? Why?

_____

_____

_____

_____

_____

Abc **Drawing Conclusions**

2. What grows on mountains below the timberline?

_____

_____

_____

⚙ **Making Predictions**

3. Could you drive from the Atlantic coastline to the Pacific coastline? Could you drive from the Pacific coastline to Hawaii? Explain your answers.

_____

_____

_____

_____

_____

_____

networks

The United States West of the Mississippi River

Lesson 1: Physical Features, *continued*

Marking the Text

4. Read the text on the right. Circle the state with the largest land area. Underline the tallest mountain in the United States.

Reading Progress Check

5. What are two ways in which the states west of the Mississippi River are similar to each other?

_____

_____

_____

_____

? Explaining

6. Why does the depth of the Great Salt Lake change?

_____

_____

Reading Progress Check

7. What landform is crossed by the rivers that flow from the Rocky Mountains to the Mississippi River?

_____

North and west of the Colorado Plateau is the Basin and Range region. This name refers to a pattern on the land of steep, high mountains separated by low-lying basins. North of the Basin and Range region is the Columbia Basin. Much of this area is flat.

Alaska is the largest state, having the most land area. Alaska is west of Canada. Mountains run along its southern and northern edges. Mount McKinley in Alaska is the highest point in the United States. It is part of the Alaska Range.

Hawaii is in the Pacific Ocean. It includes more than 130 islands. Volcanoes formed these islands. Two volcanoes in Hawaii are still active. The wind and sea have eroded some volcanic mountains. Some islands have sandy beaches along the shore.

Bodies of Water

Guiding Question *How do the bodies of water in the region affect people's lives?*

The Pacific Ocean borders the western shores of the United States. There are many major ports on the Pacific Coast. Louisiana and Texas border the Gulf of Mexico and also have many ports.

Major Port Cities West of the Mississippi River	
Pacific Coast	San Diego, Long Beach, Los Angeles, California; Portland, Oregon; Seattle, Tacoma, Washington; Valdez, Alaska; Honolulu, Hawaii
Gulf of Mexico	New Orleans, Louisiana; Houston, Texas

The coastlines of the Pacific Ocean and the Gulf of Mexico, however, are very different. The Pacific Coast has a steep coastline and rocky cliffs. The Gulf Coast is more stable and flat, with many shallow, marshy areas.

There are relatively few lakes west of the Mississippi River because of the dry climate. The largest lake west of the Mississippi River is Utah's Great Salt Lake. This is the largest salt lake in the Americas. The depth of this lake increases or decreases depending on how much water evaporates from it. Two other important lakes in the region are Lake Tahoe and Lake Mead.

There are several major rivers in the western United States. The Colorado River flows from the western slope of the Rocky Mountains to the Pacific Ocean. It provides water for farming and cities. Dams on the Columbia River provide hydroelectric power.

Lesson 1: Physical Features, *continued*

Some rivers in the west flow west toward the Pacific Ocean. Other rivers flow east toward the Gulf of Mexico. An imaginary line through the Rocky Mountains called the **Continental Divide** separates the two sets of rivers.

Climates of the Region

Guiding Question *What factors influence the climates of the region?*

The area west of the Mississippi River has many different climates, from rain forests in the Northwest to dry hot deserts in the Southwest. Mountains play an important role in forming climates. West-facing slopes of the mountains along the coastline receive a lot of rain and snow from storms coming from the Pacific Ocean. This marine west coast climate is home to vast forests.

The valleys to the east of the coastal ranges lie in the rain shadow, so they are dry. In these areas, rainwater from the mountains is used to water crops, a process called **irrigation**. The dry climate is what causes so much evaporation from Great Salt Lake. Some areas are so dry that they are covered by large deserts.

The western Great Plains have a semiarid climate. A semiarid climate has more rain than a desert but not enough for trees to grow. Temperatures get hot in the summer and cold in the winter. Sometimes in the winter a dry wind called the chinook blows.

The eastern Great Plains has two types of climates. The north has a humid continental climate because cold air masses swoop down from the Arctic. The south has a humid subtropical climate because warm, moist air blows in from the Gulf of Mexico.

The Alaskan climate is cold. Hawaii has a tropical rain forest climate with high temperatures and lots of rainfall. The table below shows the climate by region.

Region	Climate
Coastal mountains, west-facing	Marine west coast climate
Valleys east of the coastal ranges	Dry
Western Great Plains	Semi-arid climate
Eastern Great Plains (north)	Humid continental climate
Eastern Great Plains (south)	Humid subtropical climate
Alaska	Cold
Hawaii	Tropical rain forest

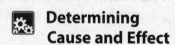 **Analyzing**

8. In what part of the country west of the Mississippi River would you like to live? Why?

_____

_____

_____

_____

_____

Marking the Text

9. Read the text on the left. Highlight areas with dry climates in yellow. Highlight areas with a wet climate in green.

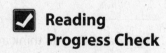 **Determining Cause and Effect**

10. What causes the south to have a humid subtropical climate?

_____

_____

✔ **Reading Progress Check**

11. Why does so much water evaporate from Great Salt Lake?

_____

_____

The United States West of the Mississippi River

Lesson 1: Physical Features, *continued*

✓ Reading Progress Check

12. Does this region of the United States rely too much on one energy resource? Why or why not?

_____

_____

_____

_____

_____

Resources of the Region

Guiding Question *What resources does the region have?*

The United States west of the Mississippi River has a variety of natural resources. These include:

- petroleum, found in the Gulf of Mexico, southern Great Plains, California, and Alaska
- natural gas and coal
- hydroelectric power produced by dams on large rivers
- wind and solar power
- **ethanol**, a fuel made from corn blended with gasoline
- gold, silver, copper, zinc and lead mined in the Rocky Mountains
- great natural beauty, which can be enjoyed in **national parks**

Writing

Check for Understanding

1. **Informative/Narrative** Choose a place west of the Mississippi River and write a paragraph describing what life might be like there. Include details about landforms and climate.

_____

_____

_____

_____

_____

2. **Argument** Do you think a region's natural beauty is a resource? Why or why not?

_____

_____

_____

_____

netw rks

The United States West of the Mississippi River

Lesson 2: History of the Region

ESSENTIAL QUESTION
How do people make economic choices?

Terms to Know

nomadic describes a way of life in which a person or group lives by moving from place to place

pueblo a village of the Pueblo people in the American Southwest

mission a Spanish church-based settlement in the west

frontier a region just beyond the edge of a settled area

Manifest Destiny the idea that it was the right of Americans to expand westward to the Pacific Ocean

annex to declare ownership of an area

extinct describing a particular kind of plant or animal that has disappeared completely from Earth

reservation an area of land that has been set aside for Native Americans

---

When did it happen?

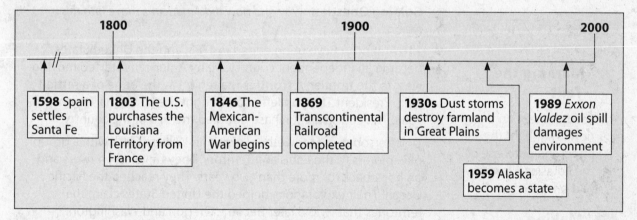

| 1800 | 1900 | 2000 |

1598 Spain settles Santa Fe

1803 The U.S. purchases the Louisiana Territory from France

1846 The Mexican-American War begins

1869 Transcontinental Railroad completed

1930s Dust storms destroy farmland in Great Plains

1989 *Exxon Valdez* oil spill damages environment

1959 Alaska becomes a state

Aᵇ𝒸 Defining

1. What does *nomadic* mean?

   _____

   _____

   _____

   _____

Early Settlements

Guiding Question *How did life in the region change for Native Americans?*

Many different Native American groups lived west of the Mississippi before Europeans arrived. Each group had its own way of life. Some settled along rivers, where they farmed and hunted. Others lived on the Great Plains and hunted bison. These groups were **nomadic**. Nomads are people who move often. Some Native American groups built teepees using animal skins and long poles. These could be folded up when the group moved.

The United States West of the Mississippi River

Lesson 2: History of the Region, *continued*

Marking the Text

2. Underline two ways the lives of Native Americans were changed when the Europeans arrived.

Reading Progress Check

3. What was the Native American way of life like before Europeans came to the region?

_____

_____

_____

_____

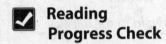
Marking the Text

4. Read the text on the right. Highlight the effect the purchase of the Louisiana Territory had on the country.

Making Connections

5. How did the belief in Manifest Destiny help people settle the West?

_____

_____

_____

_____

The Pueblo people of the Southwest lived in villages that the Spanish called **pueblos**. Pueblo means "town" in Spanish. This group built homes made of dried mud. They farmed corn, beans, and squash. In the Northwest, Native Americans fished for food. They hunted for small game on land. The thick forests in the region provided wood, which they used to build large homes.

The lives of Native Americans changed when Europeans arrived. Europeans brought new animals, such as horses, to the Americas. However, they also brought diseases that killed many native people.

Spain settled parts of California and Texas by the 1700s. Some of the settlements were **missions**, church-based towns. Spanish missionaries tried to get Native Americans to become Catholics. France claimed the land from the Rocky Mountains to the Mississippi River. This area was called the Louisiana Territory. New Orleans was the most important French town. It was founded at the mouth of the Mississippi River in the early 1700s.

Westward Expansion

Guiding Question *Why and how did Americans move into this region?*

The American Revolution ended in 1783, and the United States became an independent country. Many Americans started moving west to the frontier. A **frontier** is a region at the edge of a settled area. President Thomas Jefferson bought the Louisiana Territory in 1803. The Louisiana Purchase doubled the size of the country.

Jefferson sent Meriwether Lewis and William Clark with a group of explorers to the Louisiana Territory. They traveled on rivers and on horseback for more than two years. They reached the Pacific Ocean. Their explorations helped the United States claim the territories that would later become Oregon and Washington.

Some Americans believed in the idea of **Manifest Destiny**. They thought that the United States had a right to control all of North America, from the Atlantic Ocean to the Pacific Ocean. As a result, more and more settlers moved west. Some moved to Texas, which was part of Mexico. Texas eventually declared independence from Mexico. Then in 1845, the United States **annexed**, or took control of, Texas. That is called the Texas Annexation.

Settlement of the west caused conflict with other countries. In 1846 the United States made an agreement with Great Britain. This agreement gave the United States control of the Oregon Country, what would become the states of Washington and Oregon.

The United States West of the Mississippi River

Lesson 2: History of the Region, *continued*

War with Mexico broke out over lands in the southwest. The Treaty of Guadalupe Hidalgo in 1848 gave the United States a huge area in that region. That transfer of land was called the Mexican Cession.

Gold was found in California in 1848. Many Americans hurried there in what was called the California Gold Rush. Wherever other valuable minerals were discovered, Americans moved there in large numbers hoping to become rich. Railroads soon brought even more settlers to the region.

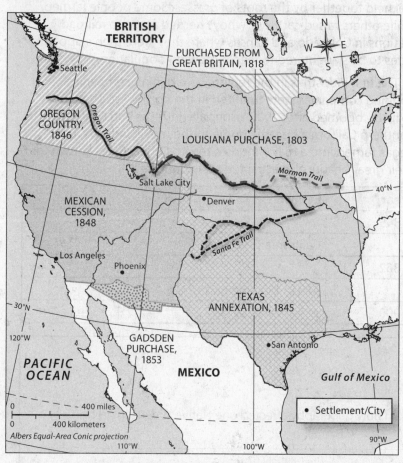

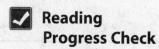

The settlers had an effect on Native Americans. Animals that the Native Americans hunted for food became **extinct**, or disappeared from Earth. The government forced many Native Americans to live on **reservations**. Reservations are lands set aside for Native Americans. These lands were usually located on poor farmland.

As settlers moved west, the United States bought Alaska from Russia in 1867. Then in 1900 the United States seized Hawaii, ignoring the Hawaiian royal family that ruled the islands.

Identifying Point of View

6. Based on the map to the left, why do you think Great Britain felt they had a claim to the Oregon Country?

_____

_____

_____

Marking the Text

7. On the map to the left, highlight the Louisiana Purchase in pink. Highlight the Oregon Country in green. Highlight the Texas Annexation in blue. Highlight the Mexican Cession in yellow.

Reading Progress Check

8. How were the acquisitions of Texas and Hawaii similar?

_____

_____

_____

_____

_____

_____

networks

The United States West of the Mississippi River

Lesson 2: History of the Region, *continued*

Marking the Text

9. Highlight the early industries that developed in the region.

Marking the Text

10. Fill out the chart on the right with important dates and events.

Reading Progress Check

11. What resources attracted Americans to the western region in the 1800s?

_____

_____

_____

Agriculture and Industry

Guiding Question *How did people in the states west of the Mississippi live?*

In 1862 Congress passed the Homestead Act. This law made land in the states west of the Mississippi River free to anyone who built a farm and lived on it for five years. Hundreds of thousands of people settled on the Great Plains to start farms.

Farm life was not easy. People built homes with sod, chunks of soil held together by the roots of grasses. Some people farmed, while others raised cattle. Cowboys herded cattle to railroad stations in Colorado and Kansas where they were then shipped to cities to be slaughtered to feed a growing population.

The first industries in the region were mining companies. Logging companies were founded in the northwest. The oil industry boomed after 1900, especially in Texas and California. Southern California became home to the movie industry. Las Vegas turned into a major tourism resort. Other places known for their great beauty also attracted many tourists.

Year	Event
1845	
	California Gold Rush
1862	
	Alaska became part of U.S.
1900	

Writing
Check for Understanding

1. **Informative/Explanatory** Explain how geography affected one Native American group.

_____

_____

2. **Informative/Explanatory** Had you lived in 1862, do you think you would have moved to the frontier to claim a homestead? Why or why not?

_____

_____

The United States West of the Mississippi River

Lesson 3: Life in the United States West of the Mississippi

ESSENTIAL QUESTION
How does technology change the way people live?

Terms to Know

Mormon a member of the Church of Jesus Christ of Latter Day Saints

Dust Bowl the southern Great Plains during the severe drought of the 1930s

topsoil the fertile soil that crops depend on to grow

agribusiness an industry based on huge farms that rely on machines and mass production methods

aerospace the industry that makes vehicles that travel in the air and in outer space

---

What do you know?

What I Know	What I Want to Know	What I Learned	How Can I Learn More

networks

The United States West of the Mississippi River

Lesson 3: Life in the United States West of the Mississippi, *continued*

Identifying Cause and Effect

1. What is one reason why western cities may not have developed good public transportation systems?

_____

_____

Reading Progress Check

2. In what ways are Las Vegas and Phoenix different from Los Angeles and Seattle?

_____

_____

_____

Marking the Text

3. Read the text on the right. Underline three groups of people who are important parts of the population in the region west of the Mississippi.

Interpreting

4. Why was NAFTA important?

_____

_____

_____

The Region's Cities and Rural Areas

Guiding Question *Where do the people of the region live?*

People who live in the western United States depend on cars to get them places. Cities are very spread out and there are large distances between them.

Many cities in the region are large ports. Los Angeles and Long Beach in California share one very important port on the Pacific Ocean. Seattle and Tacoma in Washington link the United States and Asia. Texas has three port cities on the Gulf of Mexico.

Not all busy cities are located on the coast. For example, Denver, Colorado was founded as a mining town. Today its economy is based on software, finance, and communications. Some of the most rapidly growing cities are in the interior of the United States, including Phoenix, Arizona, and Las Vegas, New Mexico.

There are many people who live in small towns and villages. Farmers, ranchers, fishermen, and others rely on the rich natural resources of the region. They may live far from cities, but use modern technology to make their lives easier.

Challenges Facing the Region

Guiding Question *What issues face the region in coming years?*

The region west of the Mississippi River is becoming more diverse. There are more Asian Americans than non-Hispanic whites living in Hawaii. Latinos outnumber non-Hispanic whites in California, New Mexico, and Texas.

The area is also home to many religious groups. For example, Utah has many **Mormons,** members of the Church of Jesus Christ of Latter Day Saints. This group settled in Utah in the mid-1800s.

The number of people older than age 65 is growing. Older people draw retirement from the government-funded Social Security system. They also need healthcare. This is usually provided through Medicare, a government-run program. Meeting the costs of these programs will be a challenge in the coming years.

Trade relations between Canada, Mexico, and the United States are strong. These countries signed a trade agreement called the North American Free Trade Agreement (NAFTA) in the early 1990s. This agreement created a free trade zone. Canada and Mexico are now the largest markets for U.S. goods. The United States also imports many things from both countries. Much of this trade flows through port cities west of the Mississippi River.

The United States West of the Mississippi River

Lesson 3: Life in the United States West of the Mississippi, *continued*

These open borders that allow goods to pass freely between countries has also made it possible for people to freely enter the country, sometimes illegally. The United States has tried to stop the flow of illegal immigrants. These efforts have had some success. A better economy in Mexico has also played a part, because Mexicans are able to get better jobs at home.

The region west of the Mississippi River faces many environmental challenges, some of them caused by nature and some caused by humans actions.

Environmental Challenges	
Natural Disasters	**Human-caused Disasters**
• drought	• oil spills
• volcanoes	• erosion
• earthquakes	• mud slides

Most of the United States west of the Mississippi has a very dry climate with little rainfall. In the 1930s a severe drought caused crops to fail in the Great Plains. The **topsoil,** the fertile soil crops need to grow, turned to dust and blew away. This area came to be called the **Dust Bowl**. Many farmers gave up farming and left the area to find other work.

New farming practices have developed that may avoid another Dust Bowl. However, as more people move to the region, more water is needed. Farmers expanding their operations also need more water. A lack of water may affect the region's economy.

Other natural disasters occur as well. Washington, Alaska, and Hawaii all have active volcanoes that can erupt. Wildfires can burn millions of acres of trees in a year. A major fault system called the **San Andreas Fault** cuts through western California. It has caused several large earthquakes. Another earthquake could cause major damage in cities such as San Francisco and Los Angeles.

There are other environmental problems that are caused by human actions. In 1989 an oil tanker spilled more than 250,000 barrels of oil into the sea off Alaska. In the Deepwater Horizon disaster of 2010 dumped nearly 5 million barrels of oil into the Gulf of Mexico. Both spills hurt wildlife and the local economy.

Erosion is caused by harvesting too many trees for lumber, leaving hillsides and mountains bare. With no roots to hold it in place, soil runs off with rain and the surface of the mountain crumbles. Very heavy rains can cause mud slides, uprooting trees and burying houses.

Identifying Cause and Effect

5. What is one negative effect of free trade?

_____

_____

_____

Making Connections

6. Why was the southern Great Plains called the "Dust Bowl"?

_____

_____

_____

Marking the Text

7. Read the text on the left. Highlight the environmental challenge that would be easiest for people to prevent.

Reading Progress Check

8. Why is the likelihood of a major earthquake along the San Andreas Fault so worrisome?

_____

_____

_____

Lesson 3: Life in the United States West of the Mississippi, *continued*

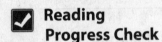
The Economy Today

Guiding Question *How do the people of the region make their living today?*

In the past, Americans moved west because of the land and natural resources. Today, most people in the west work in modern settings rather than on ranches or farms.

Farming is still important. Farms on the Great Plains produce a great deal of wheat. Other important products include cotton, corn, hay, and cattle and sheep. California grows fruits and vegetables all year. Most farms today are large agribusinesses, not small family farms. Agribusinesses rely on machines and mass-production methods to farm large areas.

Other important industries include mining, the aerospace industry, and service industries. The **aerospace** industry makes vehicles that travel in the air and in outer space. Software and information science companies have become important in some places. Large cities like Dallas and Seattle have become banking centers.

Tourism is also an important industry. Visitors travel to see the great natural beauty of the west. They enjoy unusual features like Alaska's glaciers and Hawaii's tropical beaches. National parks in the west also attract many tourists.

Writing

Check for Understanding

1. **Informative/Explanatory** What is one way that cities west of the Mississippi River differ from cities east of the river?

_____

_____

_____

2. **Narrative** Write a journal entry of a college student in New York City who has decided to move west of the Mississippi River after graduation.

_____

_____

_____

_____

Canada

Lesson 1: The Physical Geography of Canada

ESSENTIAL QUESTION

How do people adapt to their environment?

Terms to Know

province an administrative unit similar to a state
territory land administered by the national government
shield a large area of relatively flat land composed of ancient, hard rock
coniferous evergreen trees that produce cones to hold seeds and have needles instead of leaves
deciduous trees that shed their leaves in the autumn
archipelago a group of islands
tundra a flat, treeless plain with permanently frozen ground
fishery an area where fish come to feed in huge numbers

Where in the World: Regions of Canada

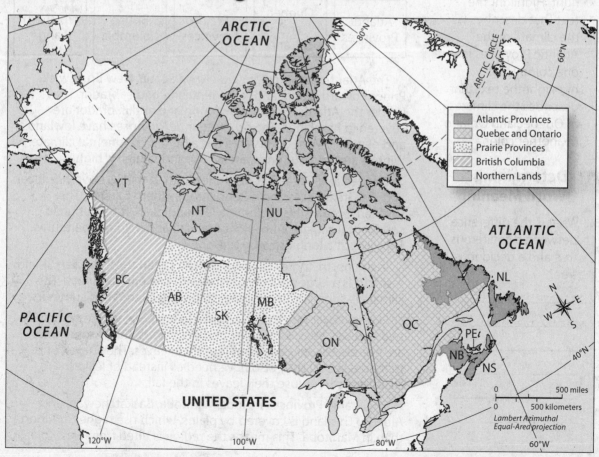

netw⊛rks

Canada

Lesson 1: The Physical Geography of Canada, *continued*

Canada's Physical Landscape

Guiding Question *How is Canada's physical geography similar to and different from that of the United States?*

Canada is a very large country with an area of 3.86 million square miles (10 million sq. km). However, it also has a low population. Most people live in the southern part of the country.

Canada is divided into 10 smaller units , or parts, called provinces. A **province** is an administrative unit similar to a state. Administrative means "managed" or "governed." Each province has its own government. Canada also has three territories. **Territories** are lands administered by the national government.

Canada is divided into five regions, as shown below.

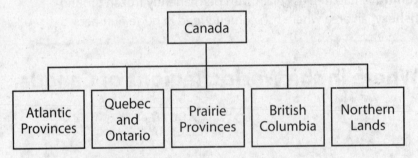

The Atlantic Provinces include Nova Scotia, New Brunswick, Prince Edward Island, and Newfoundland and Labrador. They border the Atlantic Ocean. Newfoundland and Labrador are covered in highlands. The other Atlantic Provinces have lowlands and plateaus. Most of this area has a humid continental climate. Winters are cold, summers are warm, and rainfall is high.

Quebec and Ontario start in the south at the St. Lawrence River and the Great Lakes and extend into northern Canada. The southern area of this region is a lowland plain. It has a humid continental climate with a lot of rainfall and mild temperatures. This gives it a long growing season.

In the north, there is a large plateau called the Canadian Shield. A **shield** is a large area of flat land with ancient, hard rock. The climate is subarctic. It has long, cold winters and mild summers.

Forests cover much of the northern area. These forests include two types of trees: coniferous and deciduous. **Coniferous** trees are evergreens, which do not lose their leaves. They have cones that hold seeds. They also have needles instead of leaves. **Deciduous** trees lose their leaves in the fall.

The Prairie Provinces include Manitoba, Saskatchewan, and Alberta. This land is covered by plains, which rise higher in Alberta than in Manitoba. This means the region is tilted from west to east.

Defining

1. Define *province* and *territory*.

_____

_____

_____

_____

_____

_____

Marking the Text

2. Read the text on the right. Highlight the text that describes the climate of the Atlantic Provinces in one color. Then highlight the text that describes the climate of Quebec and Ontario in another color.

Determining Word Meaning

3. What is the difference between a coniferous tree and a deciduous tree?

_____

_____

_____

_____

_____

_____

Canada

Lesson 1: The Physical Geography of Canada, *continued*

The northeastern and western areas of this region have highlands. In Manitoba, the highlands are part of the Canadian Shield. In Alberta, they are part of the Canadian Rockies.

Cold Arctic air from the north blows across this region. Winters are cold. The average temperature in January is –1° F (–18° C). Summers are much warmer. The average temperature in July is 67° F (19° C). Precipitation in this area is low.

British Columbia has the Rocky Mountains on its eastern side and the Coast Mountains on its western side. In between sits a deep valley and plateaus. The Canadian Rocky Mountains have 30 mountains that are over 10,000 feet (3,048 m) high. The Coast Mountains rise as high as 15,000 feet (4,572 m).

British Columbia has a marine west coast climate. Warm air blows west from the ocean. This means temperatures are mild. Rainfall is heavy in some areas.

The Northern Lands include Nunavut, Northwest Territories, and Yukon Territory. They all lie in northern Canada next to the Arctic Ocean. A group of islands, called an **archipelago**, sits in the Arctic Ocean. It has about 1,000 islands. Some islands are tiny, and some are very large. For instance, Baffin Island is larger than the state of California.

The center of this region is covered by lowland plains. The Canadian Shield is in the eastern side of the region. It continues north onto the islands north of Hudson Bay.

The western side of the Far North has high mountains. One of them is Mount Logan. At 19,524 feet (5,951 m), it is Canada's highest mountain.

Much of the Far North has a subarctic climate. This means it has cold winters and mild summers. The areas farthest north have a tundra climate. A **tundra** is a flat, treeless plain with permanently frozen ground.

Bodies of Water

Guiding Question *What bodies of water are important to Canada?*

Three oceans border Canada. They are the Atlantic, the Pacific, and the Arctic. They affect climate in different ways.

Ocean	Effect on Climate
Atlantic	Brings moderate temperatures in eastern Canada
Pacific	Brings rain and mild temperatures to western Canada
Arctic	Chills northern and central Canada

? Explaining

4. What is the effect of the marine west coast climate on British Columbia?

_____

_____

_____

Aᵇc Defining

5. What is an archipelago?

_____

_____

_____

✏ Marking the Text

6. Underline every time the Canadian Shield is mentioned in the text. In which regions is the shield located?

_____

_____

✔ Reading Progress Check

7. Why do most of Canada's people live in southern Canada?

_____

_____

_____

_____

Canada

Lesson 1: The Physical Geography of Canada, *continued*

Marking the Text

8. In the text, highlight the two main reasons the Atlantic and Pacific Oceans are important to Canada.

Reading Progress Check

9. Why is the St. Lawrence River economically more important to Canada than the Mackenzie River?

_____

_____

The Atlantic and Pacific Oceans are important to Canada. One reason is because ships use them to bring goods to and from Canada. Shipping is a major part of Canada's economy.

Another reason that these oceans are important to Canada is that they both have fisheries. A **fishery** is an area where fish come to feed in huge numbers. These fisheries are used by people from all over the world. However, taking too many fish from these areas has severely hurt the fish population in recent years.

The Gulf of St. Lawrence is another important body of water. It connects to the Atlantic Ocean. Ships use it to travel into Canada's interior. But without the St. Lawrence Seaway, they couldn't get all the way to western Lake Superior. The St. Lawrence Seaway is a system of canals and locks that raises and lowers ships from one height to another. It was built in the 1950s by the United States and Canada working together.

Canada shares four of the five Great Lakes with the United States. It does not share Lake Michigan. Three other major lakes are found in lowland areas west of the Canadian Shield. They are Great Slave Lake and Great Bear Lake in the Northwest Territories, and Lake Winnipeg in Manitoba.

Another important river is the Mackenzie River. With its tributaries, it flows through much of the Far North. It starts at the Great Slave Lake and empties into the Arctic Ocean.

Writing

Check for Understanding

1. **Informative/Explanatory** Select one of Canada's regions and write a paragraph describing the landforms found there and how they affect the climate.

_____

_____

_____

2. **Informative/Explanatory** Which body of water do you think is most important to Canada and why?

_____

_____

_____

networks

Canada

Lesson 2: The History of Canada

ESSENTIAL QUESTION
What makes a culture unique?

Terms to Know
aboriginal a native people
métis the children of French and native peoples
transcontinental describing something that crosses a continent
granary a building used to store harvested grain

When did it happen?

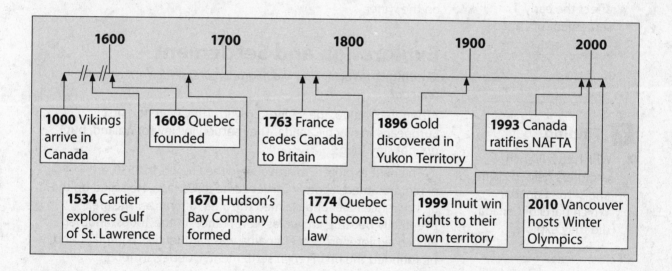

1600   1700   1800   1900   2000

1000 Vikings arrive in Canada

1608 Quebec founded

1763 France cedes Canada to Britain

1896 Gold discovered in Yukon Territory

1993 Canada ratifies NAFTA

1534 Cartier explores Gulf of St. Lawrence

1670 Hudson's Bay Company formed

1774 Quebec Act becomes law

1999 Inuit win rights to their own territory

2010 Vancouver hosts Winter Olympics

Determining Word Meaning

1. What is the definition of aboriginal?

_____

_____

The First Nations of Canada

Guiding Question *How did native peoples of Canada live before Europeans came to the area?*

The first people to live in Canada are called the First Nations. They are **aboriginal** people, or natives, who lived in North America before Europeans arrived. The aboriginals came from Asia during a period of cold called the Ice Age. The Ice Age affected where the first people lived.

Canada

Lesson 2: The History of Canada, *continued*

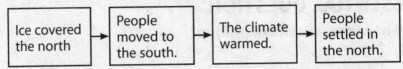

| Ice covered the north | → | People moved to the south. | → | The climate warmed. | → | People settled in the north. |

Marking the Text

2. Find the text that describes where aboriginal peoples settled after the Ice Age. Underline all three regions.

✓ Reading Progress Check

3. How did the presence and absence of ice affect the early settlement of Canada?

_____

_____

_____

? Finding

4. Which two French explorers helped claim and settle Canada? Which parts of Canada did they explore?

_____

_____

_____

⚙ Comparing

5. Compare and contrast the goals of the fur traders and the priests.

_____

_____

_____

After the Ice Age, some aboriginal people, including the Huron and Iroquois, settled in the eastern woodlands. They farmed, hunted, and fished. They also built villages and traded with one another.

Other aboriginal people lived in the west, along the Pacific Ocean. They made canoes and fished in the ocean and in rivers. They also hunted in the forests. They used the trees in the region to build their houses out of wood.

The Inuit people lived in the Far North. Few plants grow there, so they made shelters without using wood. They hunted caribou, which is an animal like a deer. They also hunted seals and whales on the water.

Exploration and Settlement

Guiding Question *How did migration and settlement change Canada?*

Around A.D. 1000, the Vikings arrived. They were the first Europeans to reach Canada. They settled in Newfoundland, but did not stay long.

The next explorers to arrive were the French. In 1530, an explorer named Jacques Cartier came to Canada. He sailed up the St. Lawrence River and claimed the area for France. The area eventually became known as New France. In the 1600s, another French explorer arrived. His name was Samuel de Champlain, and he founded the first French settlement, Quebec, in 1608.

Over time, more French settlers migrated to Canada. Some became fur traders. They traded European goods to the Huron, a First Nation's people, in exchange for beaver furs. They sent the furs back to Europe.

Other settlers were farmers and priests. Farmers grew crops that fed other settlers. Priests came to Canada to minister to the French people who were Roman Catholic. They also came because they hoped to convert native peoples to Christianity.

In the 1600s and 1700s, France was a powerful nation. But soon Britain began to compete with France for the Americas. The British formed the Hudson Bay Company. They set up trading posts in Hudson Bay in order to gain some of the fur trade.

Canada

Lesson 2: The History of Canada, *continued*

In the 1700s, Britain and France fought wars. When Britain won a victory in 1763, France had to give up much of its land in North America. However, the British passed the Quebec Act. With this Act, Britain allowed French settlers in Canada to keep their language, religion, and system of laws.

During the American Revolution, many Americans moved to Canada. Later, in 1867, the British colonies in Canada were worried that the United States would try to take over Canada. They united to become the Dominion of Canada. This new nation was partly self-governing within the British Empire.

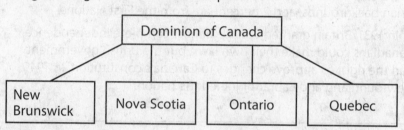

Canada's leaders wanted Canada to expand all the way to the Pacific. In 1869, Canada gained the territory held by the Hudson Bay Company. Many **métis**, who are the children of French and native peoples, lived on some of this land. The province of Manitoba was created for them.

In 1871, British Columbia on the west coast agreed to join Canada. A **transcontinental** railroad was built to link eastern and western Canada. Transcontinental means continent-crossing.

Canada's leaders also made agreements with some native peoples of the west. There was a disagreement over the use of this land, and settlers pushed out the native peoples. In 1905, Saskatchewan and Alberta joined Canada.

Meanwhile, gold had been discovered in the Yukon Territory. A gold rush caused problems, so a police force was formed in 1873. Today this force is known as the Royal Canadian Mounted Police.

Canada Grows and Unites

Guiding Question *How did Canada change in the 1900s?*

In the early 1900s, Canada's economy had problems. It was based on growing food and mining, but prices for these products fell. In response, Canada became an industrial nation. Canadians built factories and used their mineral resources. They developed

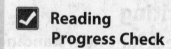

Defining

6. Who are the métis?

_____

_____

_____

Marking the Text

7. Find the text that describes how Manitoba, Saskatchewan, and Alberta joined Canada. Circle each passage.

 ### Reading Progress Check

8. How did European rivalries affect the development of Canada?

_____

_____

_____

_____

_____

Marking the Text

9. Underline the passage in the text that answers this question: Why did Canada become an industrial nation?

networks

Canada

Lesson 2: The History of Canada, *continued*

✔️ **Reading Progress Check**

10. What are two ways that Canada changed in the 1900s?

_____

_____

_____

_____

hydroelectric projects and transportation systems. After World War II, industry boomed. Agriculture grew, particularly in the west. To help feed the nation, **granaries** were used to store harvested wheat. The St. Lawrence Seaway made it easier to ship products around the world.

Canada needed more workers, so its leaders made it easier for people to enter the country. Canada's population jumped from 12 million people in 1945 to nearly 35 million in 2012. Today about half the population comes from Britain and France. Another 15 percent comes from Europe. About 6 percent have African or Asian backgrounds, and 2 percent are from the First Nations.

In 1931, Britain granted Canada almost complete independence. Canadians could make their own laws, but the British government had the right to approve changes to Canada's constitution. In 1949 Newfoundland and Labrador joined the nation.

Writing
Check for Understanding

1. Informative/Explanatory Why do you think Britain passed the Quebec Act?

_____

_____

_____

_____

2. Informative/Explanatory Write a summary that highlights the key points of Canada's development as a nation.

_____

_____

_____

_____

_____

_____

Canada

Lesson 3: Life in Canada

ESSENTIAL QUESTION

What makes a culture unique?

Terms to Know

metropolitan area an area that includes a city and the surrounding suburbs
bilingual having two official languages
peacekeeping sending the military to crisis spots to maintain peace and order
separatist a group that wants to break away from control by a dominant group
autonomy having independence from another country
acid rain rain that contains harmful amounts of poisons due to pollution

What do you know?

In the diagram below, write down everything you know and then learn about Canada's challenges.

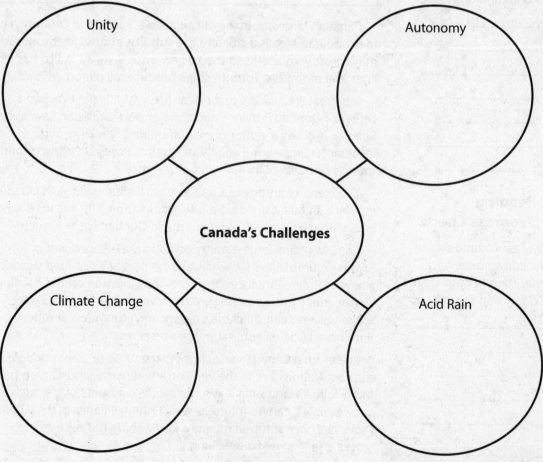

netw⭘rks

Canada

Lesson 3: Life in Canada, *continued*

Marking the Text

1. Highlight the definition of *metropolitan area*. Then provide an example of a metropolitan area that you know about.

_____

_____

Comparing and Contrasting

2. Compare and contrast Toronto and Montreal.

_____

_____

_____

_____

_____

_____

_____

Reading Progress Check

3. Why are Toronto and Vancouver more important to trade than Calgary and Edmonton?

_____

_____

_____

_____

_____

City and Country Life

Guiding Question *Where and how do Canadians live?*

Most Canadians live in cities or suburbs around cities. Ottawa is an important city in Ontario. It is Canada's capital and the home of its national government.

Canada has both a national government and regional governments in the provinces. It is like the United States in that way. Canada also has a parliamentary system, like the United Kingdom. Voters elect members of the parliament, which is Canada's legislature. The party with the most members chooses the prime minister.

The parliament and prime minister have different duties.

Government Body	Duties
Parliament	Makes laws
Prime Minister	Carries out laws

Canada's largest **metropolitan area** is Toronto in Ontario. A metropolitan area is a city and the suburbs around it. Toronto is a major port with access to the St. Lawrence Seaway. It also has rail lines and highways. Toronto ships products all over the world.

Montreal is Canada's second-largest city. It is the economic center of Quebec Province and a major port on the St. Lawrence Seaway. It is also a center of manufacturing, banking, and insurance. Quebec, the capital of Quebec Province, attracts many tourists because of its 400-year-old history.

The French heritage of Canada is particularly strong in Quebec Province. In fact, Canada is a **bilingual** nation. This means it has two official languages. Most people in Quebec speak French.

Canada's third-largest metropolitan area is Vancouver in British Columbia. It is Canada's busiest port. It ships food products from the Prairie Provinces. Because of its location on the Pacific Ocean, many products are sent from Vancouver to Asia. Other major western cities include Calgary and Edmonton in Alberta. Both have large oil and natural gas reserves.

Life in rural Canada varies from place to place. Some people of the First Nations live in the Far North where the landscape is harsh. Many follow traditional ways, but modern aspects of life, such as snowmobiles, can be found as well. Fishing villages in the Atlantic Provinces have suffered recently. Overfishing has reduced fish stocks and income from fishing.

Canada

Lesson 3: Life in Canada, *continued*

Economic and Political Relationships

Guiding Question *How does Canada get along with other nations?*

Canada has close ties to the United States. In the early 1990s, it signed a trade agreement with both the United States and Mexico. This made it easier to trade. In 2010, almost three-quarters of Canada's imports came from the United States. Also in 2010 almost three quarters of Canada's exports went to the United States.

Canada and the United States cooperate on defense. They work together to provide air support in case of attack. They also work together to fight terrorism.

Canada and the United States have cultures that are similar in many ways. For instance, Canadians watch American movies and television shows. Canadian singers and actors work in the United States. Canada is working to develop its own movie industry.

Canada is also close to Britain. Its government and laws are based on the laws in Britain (although the laws in Quebec are not). The British king or queen is Canada's king or queen, too. Canadian culture is also similar to British culture.

Canada is active in many world organizations. It helps poorer nations. It also takes part in **peacekeeping** efforts. Peacekeeping is when trained members of the military go to crisis spots. They try to keep peace and order there.

Canada's Challenges

Guiding Question *What challenges do Canadians face?*

Canada's biggest challenge might be to stay together as a nation. Some people in Quebec want to separate.

Canada's constitution makes sure that French-speaking people have rights. However, Canada is mostly controlled by English-speakers, and they have run the economy for many years. French speaking people in Quebec feel that they have been treated badly.

In the late 1990s, some Quebec leaders started a separatist movement. **Separatists** are groups who want to break away from the control of a dominant group. People in Quebec voted on this issue. They needed to decide on whether Quebec should separate from Canada. Twice they have voted no.

The First Nations people gained more power, though. In 1999, the Canadian government created a new territory for them called Nunavut. There, the people have more **autonomy**, or the ability to create their own government.

Marking the Text

4. Highlight the text that describes how Canada helps other nations.

✔ Reading Progress Check

5. Why is Canada more similar to the United States and the United Kingdom than to other nations?

_____

_____

_____

_____

Abc Determining Word Meaning

6. Use the word *separatists* in a sentence

_____

_____

_____

? Explaining

7. What is Nunavut and how is it diifferent from other territories?

_____

_____

_____

_____

Canada

Lesson 3: Life in Canada, *continued*

Another challenge Canada faces is climate change. Milder weather threatens plants and animals living in the far north. Experts believe climate change will also affect fisheries, lead to water shortages, and cause more extreme weather.

Burning fossil fuels is thought to be one cause of climate change. The Canadian government is taking some steps to reduce the use of fossils fuels. Fossil fuels include oil, coal, and natural gas.

Canada depends greatly on fossil fuels, though. They power industries, transportation, and homes. Also, a large part of Canada's economy depends on the production of fossil fuels. One of the biggest sources of oil in Canada is the Athabasca Tar Sands in northwestern Alberta.

Another environmental challenge is **acid rain**.

| Chemicals from air pollution combine with precipitation | → | Acid Rain |

When acid rain falls to Earth, it may weaken or kill fish, land animals, and trees. Damage in eastern Canada is particularly bad. Many of the chemicals that cause acid rain come from the United States. The Canadian government has made efforts to reduce acid rain. It is also trying to reach an agreement with the United States to help with this problem.

⚙ Drawing Conclusions

8. Should Canada limit fossil fuel use? Why or why not?

_____

_____

_____

☑ Reading Progress Check

9. Why is Canada unable to meet its environmental challenges by itself?

_____

_____

_____

Writing

Check for Understanding

1. **Informative/Explanatory** Why is separatism an ongoing problem in Canada?

_____

_____

_____

2. **Argument** Write an e-mail to a government leader in the United States urging strong action to cut back on air pollution that causes acid rain.

_____

_____

_____

networks

Mexico, Central America, and the Caribbean Islands

Lesson 1: Physical Geography

ESSENTIAL QUESTION
How does geography influence the way people live?

Terms to Know
isthmus a narrow strip of land that connects two larger land areas
tierra caliente the warmest climate zone, located at a lower elevation of the Tropics
tierra templada a temperate climate zone, located in a higher elevation of the Tropics
tierra fria a colder climate zone, located in a higher elevation of the Tropics
bauxite mineral ore that is used to make aluminum
extinct describing a volcano that is no longer able to erupt
dormant describing a volcano that is still capable of erupting but showing no signs of activity

Where in the world: Mexico, Central America, and the Caribbean?

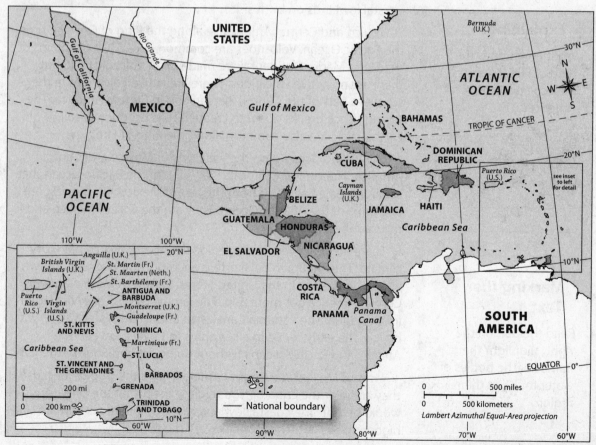

Lesson 1: Physical Geography, continued

Defining

1. What is the definition of *isthmus*?

_____

_____

_____

_____

Marking the Text

2. Read the text on the right. Highlight the names of the three mountain ranges that are found in Mexico.

Explaining

3. What effect does the Ring of Fire have on Mexico and Central America?

_____

_____

_____

_____

_____

_____

_____

Marking the Text

4. Read the text on the right. Highlight the names of the bodies of water found in the region.

Physical Geography of Mexico and Central America

Guiding Question *What landforms and waterways do Mexico and Central America have?*

Mexico and Central America form an **isthmus**, a narrow strip of land, that connects North and South America. Along with South America and some Caribbean islands, they make up Latin America. Spanish and Portuguese are spoken in this region of the Americas. Both languages are based on Latin, the language of ancient Rome.

Mexico is the largest nation of the region, with about two-thirds of the land. On the north, it has a long border with the United States. Two coastal mountain ranges form a backwards *y*. The Sierra Madre Occidental is on the west, and the Sierra Madre Oriental is in the east. There are coastal plains alongside both ranges.

Between the mountain ranges is a high central plateau. The two ranges join in the Southern Highlands, which form the tail of the *y*. Mountains run down the center of Central America, with narrow coastal lowlands on both sides.

Mexico and Central America lie along the Ring of Fire that rims the Pacific Ocean. Volcanoes are common here. The mountains of the Sierra Madre Occidental are made up of volcanic rock, but there are no active volcanoes. There are active volcanoes in the southern part of the Central Plateau and in Central America. The volcanic rock breaks down to create fertile, productive soil. Earthquakes are also common along the Ring of Fire.

The Pacific Ocean is on the west side of Mexico and Central America. The Gulf of California is an inlet of the Pacific Ocean that separates Baja California from the rest of Mexico. On the east are two arms of the Atlantic Ocean. They are the Gulf of Mexico and the Caribbean Sea.

There are few rivers in the region. Northern Mexico has a dry climate. The Rio Bravo del Norte is an important river. It is known as the Rio Grande in the United States. Southern Mexico and Central America get more rain. The largest lake is Lake Nicaragua in Nicaragua. The Panama Canal is an important waterway. It was built in the early 1900s to allow ships to travel between the Atlantic and Pacific oceans without going around South America.

Most of Mexico and Central America lie in the tropics. Because they are near the Equator, you might expect that the climate would be hot. The coastal lowlands are hot, but areas that are higher up are not. The highlands are much cooler.

Mexico, Central America, and the Caribbean Islands

Lesson 1: Physical Geography, *continued*

Geographers divide nearly all the region into three climate zones based on their elevation, or how high they are. Soil, crops, animals, and climate change from zone to zone.

The **tierra caliente** ("hot land") is the warmest zone. Tropical crops such as bananas, sugarcane, and rice grow there.

Slightly higher in elevation, the **tierra templada** ("temperate land") has a cooler climate. Coffee, corn, and wheat grow well in this zone and most of the people of the region live there.

At an even higher elevation is the **tierra fria** ("cold land"), which has chilly nights. Hardy crops such as potatoes, barley, and wheat grow there. Dairy farming is also a major agricultural activity in this climate zone.

A fourth zone above this is the *tierra helada* ("frozen land"), where few human activities take place. However, this climate zone is more common in other areas of the Americas.

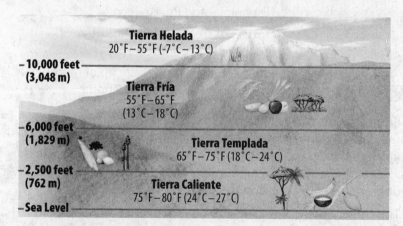

Tierra Helada
20°F–55°F (-7°C–13°C)

—10,000 feet
(3,048 m)

Tierra Fría
55°F–65°F
(13°C–18°C)

—6,000 feet
(1,829 m)

Tierra Templada
65°F–75°F (18°C–24°C)

—2,500 feet
(762 m)

Tierra Caliente
75°F–80°F (24°C–27°C)

—Sea Level

Much of Mexico and Central America has a tropical wet/dry climate. Most of the precipitation falls during the wet summer season. There is a dry winter season that is longer in areas farther from the Equator. Fierce hurricanes can strike during the summer and early fall.

Mexico's most important resources are oil and natural gas. They are found along the coast of the Gulf of Mexico and in the gulf waters. The Spanish were first attracted to the area's gold and silver. Other minerals found here include copper, iron ore, and **bauxite**, a mineral ore used to make aluminum.

Most of Central America has few mineral resources. Nicaragua is an exception. It has gold, silver, iron ore, lead, zinc, and copper. Guatemala has some oil, and its mountains produce nickel.

Drawing Conclusions

5. Why do most of the people in Mexico and Central America live in the *tierra templada* climate zone?

_____

_____

Marking the Text

6. Read the text on the left. Underline the sentences that describe a tropical wet/dry climate.

Marking the Text

7. Circle the mineral that is used to make aluminum.

Reading Progress Check

8. Why are different climate zones found in this region, even though most of the region is in the tropics?

_____

_____

_____

_____

networks

Mexico, Central America, and the Caribbean Islands

Lesson 1: Physical Geography, *continued*

Marking the Text

9. Read the text on the right. Highlight the names of the two countries that share the island of Hispaniola.

Drawing Conclusions

10. Why do you think that some of the islands of the Caribbean are called the Greater Antilles, while others are called the Lesser Antilles?

_____

_____

_____

_____

_____

_____

Defining

11. How do *extinct* volcanoes differ from active volcanoes?

_____

_____

_____

_____

_____

_____

Physical Geography of the Caribbean Islands

Guiding Question *How are the Caribbean islands alike and different from one another?*

Hundreds of islands dot the Caribbean Sea. They are home to more than 30 countries or territories belonging to other countries. Some are large, with millions of people. Others are tiny, and home to only thousands.

The Caribbean islands can be split into three groups: the Greater Antilles, the Lesser Antilles, and the Bahamas. Each group contains many islands.

Caribbean Island Groups	
Greater Antilles	Four large islands: Cuba, Jamaica, Hispaniola, and Puerto Rico
Lesser Antilles	Dozens of smaller islands, mostly independent countries
Bahamas	Independent nation made up of more than 3,000 islands east of Florida

Cuba and Jamaica are independent countries. Hispaniola is home to two countries. Haiti is located on the west side of the island, and the Dominican Republic is on the east side. Puerto Rico is a U.S. commonwealth. It has its own government, but the people are American citizens.

The Lesser Antilles were once colonies of France, Britain, Spain, or the Netherlands. Now they are independent countries. Their cultures reflect their colonial past.

The Greater Antilles are a mountain chain. Much of this mountain chain is under water. The Lesser Antilles were formed by volcanoes. Many of these volcanoes no longer erupt because they are **extinct.** However, some are only **dormant.** That means they could erupt but do not show any signs of being active.

The Caribbean Sea is a western arm of the Atlantic Ocean. Its warm waters help feed the Gulf Stream. This current carries warm water up to the eastern coast of the United States. The Caribbean islands have a tropical wet/dry climate.

Temperatures are high year-round, but ocean breezes make life comfortable. Humidity is generally high. Rainfall varies. Some islands can get only 10 inches of rain a year. Others can get as

Mexico, Central America, and the Caribbean Islands

Lesson 1: Physical Geography, *continued*

much as 350 inches a year. The islands, mostly those in the north, are prone to hurricanes.

The Caribbean Sea is rich in fish. Some are harvested for food and others are for sport fishing. The islands have little timber today, and there are few mineral resources. Some Caribbean islands have important resources, though. Trinidad and Tobago have oil and natural gas. The Dominican Republic exports nickel, gold, and silver. Cuba is a major nickel producer. Jamaica has large amounts of bauxite.

However, the islands' most important resources are their climate and people. Millions of tourists come each year to enjoy the sandy beaches and warm hospitality.

> ✓ **Reading Progress Check**
>
> 12. How did the islands of the Caribbean form?
>
> _____
>
> _____
>
> _____
>
> _____
>
> _____

Writing

Check for Understanding

1. Informative/Explanatory How is the physical geography of Mexico and Central America similar?

_____

_____

_____

_____

2. Informative/Explanatory Why do the islands of the Caribbean Sea have so many different cultures?

_____

_____

_____

_____

Mexico, Central America, and the Caribbean Islands

Lesson 2: History of the Regions

ESSENTIAL QUESTION
Why does conflict develop?

Terms to Know

staple a food that is eaten regularly

surplus extra; more than needed

conquistador a Spanish explorer of the early Americas

colonialism a policy based on control of one country by another

revolution a period of violent, sweeping change

plantation a large farm

cash crop farm product grown for sale

caudillo military strongmen who ruled the region's countries

Columbian Exchange the transfer of plants, animals, and people between Europe, Asia, and Africa on one side and the Americas on the other

---

When did it happen?

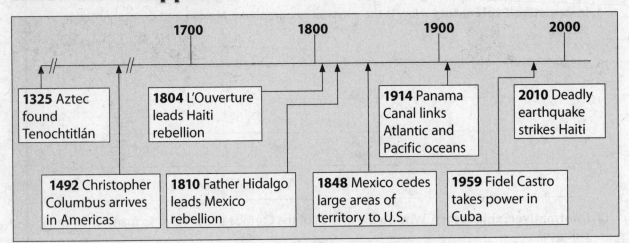

1325 Aztec found Tenochtitlán

1492 Christopher Columbus arrives in Americas

1804 L'Ouverture leads Haiti rebellion

1810 Father Hidalgo leads Mexico rebellion

1848 Mexico cedes large areas of territory to U.S.

1914 Panama Canal links Atlantic and Pacific oceans

1959 Fidel Castro takes power in Cuba

2010 Deadly earthquake strikes Haiti

Marking the Text

1. Underline the staple foods of the early Mexicans.

Mexico's History

Guiding Question *How did economic and governmental relationships between Spanish and Native Americans in Mexico change over time?*

Native peoples first grew corn in Mexico about 7,000 years ago. They also grew other **staples**, or foods eaten regularly, such as squash, chilies, and avocados. They were able to produce more food than they needed to survive. This **surplus**, or extra food, allowed them to specialize in other jobs.

networks

Mexico, Central America, and the Caribbean Islands

Lesson 2: History of the Regions, *continued*

The Maya formed a civilization about 3,000 years ago. They lived mainly in Mexico's Yucatán Peninsula and what is now Guatemala and Belize. One feature of their culture was great cities. They built pyramids and invented a complex writing system. Their civilization reached its peak from A.D. 300 to 900. Then it mysteriously collapsed.

The Aztec settled in central Mexico in about 1300. Their capital was Tenochtitlán, where Mexico City now stands. The Aztec had a complex social and religious system. They conquered many of their neighbors and made slaves of captured soldiers. They were also skilled farmers.

Around 1520, Hernán Cortés led a small force of Spanish **conquistadors**, or conquerors, to Mexico. Within two years, they had taken control of the Aztec empire. They were able to do this because their guns and armor were superior to Aztec spears. The Spanish also carried diseases for which the Native Americans had no resistance. These diseases killed many thousands. Many native peoples who resented Aztec rule joined Cortés as allies.

The conquest completely changed life in Mexico. It brought Spain riches in gold and silver mines. Priests converted native peoples to the Catholic faith. Native peoples were forced to work on farms or in mines. Spanish rule was an example of **colonialism**. Under colonialism, one nation controls an area and its government, economy, and society. The colony's resources go to make the ruling nation rich. Settlers from Spain controlled most of Mexico's wealth.

In 1821, Mexico gained its independence after 300 years of Spanish rule. However, little changed for the people who lived there. The country was ruled by wealthy landowners and native peoples remained poor. In the early 1900s, a **revolution** broke out. A revolution is a sweeping change. In Mexico, there was a new constitution and land reform. Land was divided into parcels given to poor people to farm.

A History of Central America

Guiding Question *How did the nations of Central America develop?*

The nations of Central America developed in similar ways to Mexico. However, there were also differences. The Maya thrived in Guatemala and Belize, as well as southern Mexico. Even after they abandoned their cities, they continued to live in the region. After conquering Mexico, the Spanish moved south. By the 1560s Spain controlled most of Central America. In the early 1800s, Britain claimed the area that is now Belize.

✏ Marking the Text

2. Read the text on the left. Highlight the names of native groups that ruled parts of Mexico before the Spanish.

✏ Marking the Text

3. Underline the definition of *colonialism*.

⚙ Analyzing

4. Why would Mexico's fight for independence during the 1800s not necessarily be considered a revolution?

_____
_____
_____
_____
_____

✔ Reading Progress Check

5. How were the Spanish able to conquer the Aztec?

_____
_____
_____
_____

netw⊙rks

Mexico, Central America, and the Caribbean Islands

Lesson 2: History of the Regions, *continued*

Central America became independent soon after Mexico. In 1823, Central American territories united to form one government. By 1840, they had separated into five independent countries. Belize was still a British colony, and Panama was part of Colombia.

Original Nations of Central America: Costa Rica, Nicaragua, Honduras, Guatemala, El Salvador

The countries of Central America were subjected to economic colonialism. Large companies from other countries dominated the people economically. They set up **plantations**, or large farms, to grow **cash crops**. These are crops grown for export. The people who worked on these plantations were poorly paid. The most important cash crops were bananas, coffee, and sugarcane.

During most of this time, military strongmen called **caudillos** controlled the governments. The caudillos helped ensure the foreigners' success. The foreigners made sure the caudillos stayed in power.

Around 1900, the United States helped Panama gain independence from Colombia. The United States wanted to build a canal there. The United States controlled the canal until 2000. Then, by agreement, Panama took control of the canal.

The late 1900s were a time of conflict. The upper classes became richer. Most people, however, remained poor. Various groups demanded reforms. Several countries suffered from civil wars. Only Costa Rica and Belize remained peaceful. Costa Rica's president Óscar Arias Sánchez helped bring peace to the region.

History of the Caribbean Islands

Guiding Question *How did the Caribbean islands develop?*

The history of the Caribbean islands is similar to that of Mexico and Central America. They have greater diversity because several European countries ruled them as colonies. Spain had colonies in what are now Cuba, the Dominican Republic, and Puerto Rico. The French settled what is now Haiti and some smaller islands. The British and Dutch also had colonies.

Defining

6. What is the difference between a *conquistador* and a *caudillo*?

_____

_____

Marking the Text

7. Read the text on the right. Highlight the names of cash crops that were grown in Central America.

Drawing Conclusions

8. Why do you think the United States helped Panama gain freedom from Colombia?

_____

_____

_____

Reading Progress Check

9. How did Central America and Mexico's history differ?

_____

_____

_____

_____

_____

networks

Mexico, Central America, and the Caribbean Islands

Lesson 2: History of the Regions, *continued*

In the 1600s, the Caribbean colonies became the center of the sugar industry. Many Native Americans workers died from disease, overwork, and starvation. Europeans brought in hundreds of thousands of enslaved Africans to work the plantations. This was part of the Columbian Exchange. The **Columbian Exchange** was the movement of plants, animals, and people between Europe, Asia, and Africa on one side and the Americas on the other. This transfer also introduced several new diseases around the world.

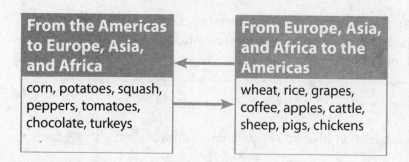

From the Americas to Europe, Asia, and Africa	From Europe, Asia, and Africa to the Americas
corn, potatoes, squash, peppers, tomatoes, chocolate, turkeys	wheat, rice, grapes, coffee, apples, cattle, sheep, pigs, chickens

In 1804 Haiti gained its independence. The Dominican Republic won independence in 1844. Cuba and Puerto Rico were Spanish until the Spanish-American War in 1898. Cuba became independent, and Puerto Rico passed to the United States. Other islands did not become independent until the middle 1900s.

Rule by caudillos and poverty have remained a problem in Haiti and the Dominican Republic. Fidel Castro took over Cuba in 1959. His communist government controlled all areas of the economy and society. Other islands have had difficulties. With few resources, they have been unable to develop strong economies. Many rely on aid from the countries that used to run them as colonies.

✎ Marking the Text

10. Read the text on the left. Cuba is the largest island nation in the Caribbean. Highlight all the references to Cuba in this section.

? Identifying

11. What was the *Columbian Exchange*?

_____

_____

✔ Reading Progress Check

12. What caused the population of the Caribbean islands to grow in colonial times?

_____

_____

_____

Writing

Check for Understanding

1. **Informative/Explanatory** How did colonialism affect the nations of Mexico, Central America, and the Caribbean?

_____

_____

_____

netw⊙rks

Mexico, Central America, and the Caribbean Islands

Lesson 3: Life in the Region

ESSENTIAL QUESTION
Why do people trade?

Terms to Know
maquiladora a foreign-owned factory where workers assemble parts
mural a large painting on a wall
dependence too much reliance
free-trade zone an area where trade barriers between countries are relaxed or lowered
remittance the money sent back home by people who have gone somewhere else to work
reggae a traditional Jamaican style of music that uses complex drum rhythms

What Do You Know?
In the first column, answer the questions based on what you know before you study. After this lesson, complete the last column.

Now...		Later...
	How much of Mexico's output comes from factories?	
	What problems do residents of Mexico City face?	
	What products are manufactured in Central America?	
	What problems does Haiti face?	

Defining

1. How does a *maquiladora* differ from a regular factory?

_____

_____

_____

_____

Modern Mexico

Guiding Question *What is life like in Mexico today?*

Mexico has a rich culture and is a rising economic power. It has close economic ties to the United States and Canada. Those three countries are joined in the North America Free Trade Agreement (NAFTA). Most of Mexico's trade is with the United States.

Factories account for about a third of Mexico's output. Some of them are **maquiladoras**, foreign-owned factories where products are assembled for export. Many are in northern Mexico. Textile and clothing industries, heavy manufacturing, and food processing are also important. Mexico produces iron, steel, and cars.

Mexico, Central America, and the Caribbean Islands

Lesson 3: Life in the Region, *continued*

Agriculture remains important. Cotton and wheat grow in the north. Along the southeastern coast, farms produce coffee, sugarcane, and fruit. Farmers grow corn, wheat, fruits, and vegetables on the central plateau. Many farmers in the poor south grow just enough to feed their families.

Service industries are also important. Banking helps economic growth. A major service industry is tourism. People come to Mexico from around the world. Some visit ancient Maya sites and tour cities with Spanish colonial architecture. Others come to relax in resorts along the tropical coasts.

Mexicans are proud of their blend of Spanish and native cultures. In the early 1900s, several Mexican painters created impressive **murals**, which are large wall paintings, to celebrate Mexico's history and people. The Ballet Folklórico performs Mexican dances. Sports reflect Mexico's ties to Spain and the United States. Both soccer and baseball are popular.

Mexico City is one of the largest cities in the world. With its suburbs, it has more than 21 million people. This is nearly 20 percent of Mexico's population. Overcrowding and pollution are both problems. Air pollution from cars and factories is sometimes held in place by mountains. The result can be a serious threat to health.

Criminals who sell illegal drugs are another challenge facing Mexico. Drug lords use violence to fight police and frighten people. Mexico has been battling this problem with some success. Poverty is another major challenge. From one-fifth to nearly half of Mexico's people are poor. The economy is growing, however, and some economists predict that Mexico will become the region's leading economy in the 2010s.

Modern Central America

Guiding Question *What is life like in Central America?*

Central America has fewer resources than Mexico. Poverty is widespread. Some of the countries of Central America have begun to escape **dependence**, or too much reliance, on cash crops. Both manufacturing and tourism have grown. Most manufacturing is food processing and clothing and textile production. Tourists visit ancient Maya sites in Belize and Guatemala. They visit Costa Rica to see the plants and animals of the rain forest.

Panama benefits economically from the Panama Canal. It has begun a project that will expand the canal so that it can handle larger cargo ships.

Marking the Text

2. Read the text on the left. Highlight the agricultural products produced in the different regions of Mexico.

❓ Identifying

3. What challenges does Mexico face?

_____

_____

_____

_____

✅ Reading Progress Check

4. How have close ties with the United States helped Mexico's economy?

_____

_____

_____

_____

Marking the Text

5. Read the text on the left. Underline the sentences that tell about tourist attractions in Central America.

Mexico, Central America, and the Caribbean Islands

Lesson 3: Life in the Region, *continued*

Copyright © The McGraw-Hill Companies, Inc.

Comparing and Contrasting

6. What similarities and differences are there between NAFTA and CAFTA-DR?

   _____

   _____

   _____

   _____

Marking the Text

7. Circle the definition of a *free-trade zone*. Why would nations want to be part of one?

   _____

   _____

   _____

Reading Progress Check

8. Why is poverty such a big problem in Central America?

   _____

   _____

   _____

Marking the Text

9. Underline the text that tells what life in Cuba is like today.

The countries need to grow their economies to provide jobs for growing populations. Trade agreements provide a way to promote growth. In the 2000s, the United States and the Dominican Republic signed agreements with five Central American countries. The Central American Free Trade Agreement (CAFTA-DR) creates a **free-trade zone.** This means that trade barriers between these countries are lowered.

United States

Dominican Republic

Costa Rica

CAFTA-DR

Honduras

El Salvador

Guatemala

Modern Central America faces several challenges. Natural disasters pose another challenge to the area. Earthquakes and hurricanes can have a serious effect on a nation's economy. The region is also held back by political problems. The civil wars of the 1980s and 1990s are over. Some of the issues that caused them still exist. Conflicts could resume.

Central America's culture is influenced by European and native traditions. Spanish is the chief language in most of the region. In Belize, English is the official language. English is also spoken in many cities in the region. Native languages are common in rural Guatemala.

Most of the population is of mixed European and native heritage. Some people of African and Asian descent also live there. Most of the people are Roman Catholics. In recent years, Protestant faiths have gained followers.

The Caribbean Islands

Guiding Question *What is life like on the Caribbean islands?*

Most of the Caribbean islands are small countries. They have small populations and few resources. The biggest challenge for most is to develop economically. There is high unemployment in Puerto Rico. Cuba's economy is in poor condition. It relies on aid from Venezuela. Under communism, Cubans have little political freedom.

Haiti has a history of poor leadership. It is one of the world's poorest nations. There is also widespread disease. As many as one in eight Haitians have left the country. Among them were Haiti's most educated people. The country has not recovered from a deadly 2010 earthquake.

networks

Mexico, Central America, and the Caribbean Islands

Lesson 3: Life in the Region, *continued*

Smaller Caribbean islands have had more political success. Governments are democratic and stable. However, their economies are plagued by few resources and poverty. **Remittances**, or money sent back by people who migrated to other lands for work, play an important part in island economies.

Tourism is a major part of the economy of several islands. There are resorts in the Bahamas, Jamaica, and other islands. Resorts provide jobs for island citizens.

Caribbean cultures show a mix of mainly European and African influences. Large numbers of Asians also came to some of the islands in the 1800s and 1900s. Languages spoken on the islands reflect their colonial heritage. English is taught in Puerto Rico's schools. Creole, a blend of French and African languages, is also spoken in Haiti.

Languages of the Caribbean Islands	
French	• Haiti
English	• Bahamas
	• Jamaica
Spanish	• Cuba
	• Dominican Republic
	• Cuba

Music of the Caribbean islands blends African and European influences. Cuba is famous for its *salsa*, and Jamaica for **reggae**. Reggae is popular around the world not only for its musical qualities but also for lyrics that protest poverty and lack of equal rights.

⚙ Drawing Conclusions

10. How do the languages of the Caribbean islands reflect their colonial history?

_____

_____

_____

_____

✓ Reading Progress Check

11. How do economic conditions in Jamaica relate to the development of reggae?

_____

_____

_____

_____

Writing

Check for Understanding

1. Informative/Explanatory List three industries that are important in Mexico.

_____

_____

2. Informative/Explanatory How have natural disasters affected the region's economies?

_____

_____

Brazil

Lesson 1: Physical Geography of Brazil

ESSENTIAL QUESTION
How does geography influence the way people live?

Terms to Know

tributary small river that flows into a larger river

basin an area into which a river and its tributaries drain

rain forest dense stand of trees and other growth that receives high amounts of precipitation

canopy umbrella-like covering formed by the tops of trees in a rain forest

plateau an area of high, flat land

escarpment steep cliff at the edge of a plateau with a lowland area below

pampas treeless grassland

Tropics area between the Tropic of Cancer and the Tropic of Capricorn, which has generally warm temperatures because it receives the direct rays of the sun for much of the year

temperate zone area between the Tropic of Capricorn and the Antarctic Circle and between the Tropic of Cancer and the Arctic Circle

Where in the World: Colonies in Brazil

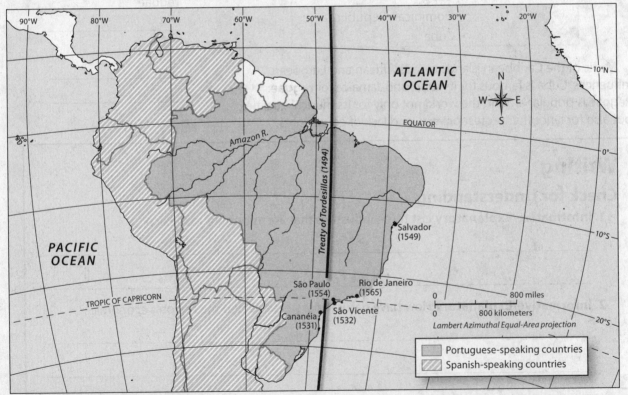

Brazil

Lesson 1: Physical Geography of Brazil, *continued*

Waterways and Landforms

Guiding Question *What are Brazil's physical features?*

The Amazon River starts in the Andes of Peru and flows east to the Atlantic Ocean. The Amazon is the world's second longest river. However, it is the largest river in the world in terms of how much freshwater it carries. One reason it carries so much water is that it has more than 1,000 smaller rivers that flow into it. These smaller rivers are called **tributaries**.

The Amazon and its tributaries empty into the Amazon Basin. A **basin** is an area into which a river and its tributaries drain. The Amazon Basin covers more than 2 million square miles (5.2 million sq. km). Much of the Amazon Basin is covered by the world's largest **rain forest**, or warm woodland with a high yearly rainfall.

In the rain forest, tall evergreen trees form a **canopy**, or an umbrella-like covering. Called the Selva, this rain forest is the world's richest biological resource. It is home to several million kinds of plants, insects, birds, and other animals.

South and east of the Amazon River are the Brazilian Highlands. This is mainly a region of rolling hills and areas of high, flat land called **plateaus**. The highlands are divided into western and eastern parts. The western part of the highlands is made up of grassland. Farming and ranching are the main activities here.

The eastern part contains low mountain ranges. In other places, highland plateaus plunge to the Atlantic coast where the land forms **escarpments**. These are steep cliffs at the edge of a plateau with a lowland area below. Farther south are grassy, treeless plains called **pampas**. Farmers and ranchers make good use of the fertile soil here. Two of Brazil's largest cities are in the highlands.

Brasilia	• Third-largest city in Brazil • Population of 3.5 million people • Capital city
São Paulo	• Largest city in Southern Hemisphere • Population of 17 million people • Important industrial city

Brazil has one of the longest strips of coastal plains in South America. It is wedged between the Brazilian Highlands and the Atlantic Ocean. This narrow plains region is called the Atlantic lowlands. The coastal lowlands cover a small part of Brazil's territory but most of the population lives there. Rio de Janeiro, Brazil's second-largest city, lies on these lowlands.

Defining

1. In the text, find the definitions of *tributary* and *basin*. Highlight them in different colors.

Drawing Conclusions

2. Why do you think the Amazon rain forest is the world's richest biological resource?

_____

_____

_____

_____

Determing Word Meaning

3. What is an escarpment?

_____

_____

_____

Reading Progress Check

4. Why do many Brazilians live in the Brazilian Highlands?

_____

_____

_____

_____

Brazil

Lesson 1: Physical Geography of Brazil, *continued*

Contrasting

5. What is the difference in location between the Tropics and the temperate zone?

_____

_____

_____

_____

Marking the Text

6. How many different climates does Brazil experience? Circle each one and write the total here.

_____

Reading Progress Check

7. What factors make farming in the northeastern part of Brazil difficult?

_____

_____

_____

Describing

8. List the two areas in Brazil where natural resources are mined.

_____

_____

A Tropical Climate

Guiding Question *What are Brazil's climate and weather like?*

Most of Brazil is located in the **Tropics**. This is the zone along Earth's Equator that lies between the Tropic of Cancer and the Tropic of Capricorn. The area along the Equator in northern Brazil has a tropical rain forest climate. Every day is warm and wet.

The area along the Amazon River also has a tropical rain forest climate. Winds called monsoons bring a huge amount of rain. In monsoon season, flooding swells the Amazon River. During dry season, forest fires are a danger.

Most of the northern and central Brazilian Highlands have a tropical wet/dry climate. As the name implies, there are two seasons—a wet summer with lots of rain and a dry winter with very little rain. Daily average temperatures are around 70°F (21°C) in the summer and 60°F (16°C) in the winter.

The northeastern part of the Brazilian Highlands is the hottest and driest part of the country. It has a semiarid climate where summer temperatures reach 100°F (37.8°C). Droughts are frequent and severe.

Southeastern Brazil is located in the **temperate zone**. This is the region between the Tropic of Capricorn and the Antarctic Circle. It has a humid subtropical climate. Summers are warm and winters are mild. Rainfall occurs year-round.

Natural Resources

Guiding Question *What resources are most plentiful and important in Brazil?*

Brazil has some of the world's most plentiful natural resources, especially in the south and southeast. Forests cover about 60 percent of Brazil. Heavy logging occurs in the Atlantic lowlands. However, logging in the Amazon Basin is increasing as more roads are built and settlement grows. The Amazon Basin contains mahogany and other hardwoods. It also houses natural rubber, nuts, and medicinal plants.

Brazil is rich in mineral resources that are only partly developed. They include iron ore, tin, copper, bauxite, gold, and manganese. At one time, most mining was done in the Brazilian Highlands, but recently, major deposits of minerals have been found in the Amazon. Brazil also has huge potential reserves of petroleum and natural gas. These reserves are found under the ocean floor off the coast.

Brazil

Lesson 1: Physical Geography of Brazil, *continued*

Brazilian farms produce food for Brazil's people as well as goods to be exported.

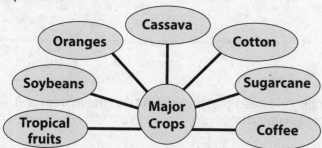

Brazil grows more coffee than any other place in the world. It is grown mostly in the eastern Brazilian Highlands and the Atlantic lowlands. Coffee used to be Brazil's main export, but today it is soybeans. These are grown mostly in the south, but also in the Brazilian Highlands.

Brazil grows one-third of the world's oranges, and the country is the world's leading supplier of citrus fruit. It is also the largest beef exporter in the world. Most of the country's grazing land is in the south and southeast.

Brazil is also a world leader in sugarcane, which can be used to make ethanol. Ethanol is mixed with gasoline and used as fuel in cars and trucks. Cars in Brazil use ethanol as fuel.

✎ Marking the Text

9. In the text, underline the products of which Brazil is the number one provider in the world.

✓ Reading Progress Check

10. Which two regions are Brazil's most important agricultural areas?

_____
_____
_____
_____

Writing

Check for Understanding

1. Informative/Explanatory What resources are important to Brazilian exports?

_____
_____
_____

2. Argument In which of Brazil's physical regions would you most like to live? Why?

_____
_____
_____
_____

Brazil

Lesson 2: History of Brazil

ESSENTIAL QUESTION

How do governments change?

Terms to Know

indigenous native

slash-and-burn agriculture a method of farming in forests that involves cutting down trees and burning away underbrush to create fields for growing crops

emancipate set free

compulsory required

When did it happen?

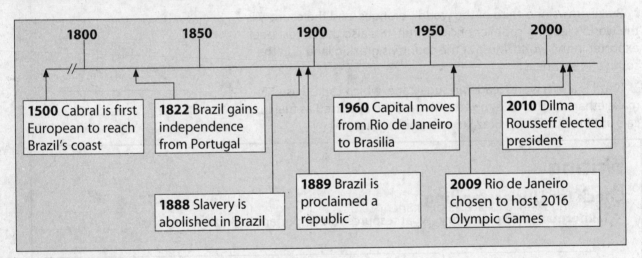

1800   1850   1900   1950   2000

1500 Cabral is first European to reach Brazil's coast

1822 Brazil gains independence from Portugal

1960 Capital moves from Rio de Janeiro to Brasilia

2010 Dilma Rousseff elected president

1888 Slavery is abolished in Brazil

1889 Brazil is proclaimed a republic

2009 Rio de Janeiro chosen to host 2016 Olympic Games

Activating Prior Knowledge

1. When Columbus arrived in the Americas, what was he looking for?

_____

_____

Early History

Guiding Question *How did Brazil's early peoples live?*

In 1493, Christopher Columbus returned to Spain with news of his explorations in the Americas. The Pope divided the new land between Spain and Portugal in 1494. Spain got a large portion of land. Portugal received just the eastern part of South America.

When the Portuguese sailed to Brazil in 1500, they met peacefully with the **indigenous**, or native, people, who lived along the coast. The Portuguese commander, Pedro Cabal, claimed the land for Portugal.

networks

Brazil

Lesson 2: History of Brazil, *continued*

Many different native peoples had lived in Brazil for more than 10,000 years. By 1500, the population had grown to between 2 million and 6 million. There were four main language groups. Each group included many native peoples.

One group of native people was the Tupi. They lived along the coast and in the rain forests south of the Amazon River. They grew cassava, corn, sweet potatoes, beans, and peanuts. They hunted fish and other water animals with arrows and harpoons from canoes.

Other native groups included the Arawak and Carib people of the northern Amazon and coast. The Nambicuara group lived in the drier grasslands and highlands. Like the Tupi, Brazil's other lowland and rain forest peoples were mainly farmers. They lived in permanent villages and governed themselves. They used **slash-and-burn agriculture**. This is a method of farming in forests that involves cutting down trees and burning away the underbrush to create fields for growing crops.

Farther south, most of the Nambicuara people were nomads who moved from place to place. In the dry season, they hunted, fished, and collected seeds, roots, and other parts of trees and plants. In the wet season, they built temporary villages and used slash-and-burn agriculture.

The Portuguese created trading posts along the coast and collected brazilwood. The red dye in this wood was valued in Europe, and it made Europeans interested in the region. Because of the trade, the Portuguese named the region Brazil. In 1533, Portugal's King John III created a permanent colony in Brazil. He wanted to maintain control of the area.

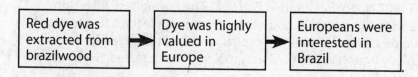

Colonial Rule

Guiding Question *How did the Portuguese colony in Brazil develop?*

King John III gave his supporters land in Brazil that extended west from the coast about 150 miles (241 km) inland. In return, the people had to develop the land. To help with their workload, the colonists enslaved nearby native peoples. Many natives resisted and were killed. Thousands more died from European diseases they had no resistance to. In 1549, King John put Brazil under royal control.

✏ Marking the Text

2. Circle the four native peoples described in the text to the left. Underline the text that describes where they lived.

🔤 Defining

3. What is slash-and-burn agriculture?

_____

_____

_____

✔ Reading Progress Check

4. Why did the Portuguese colonize Brazil?

_____

_____

_____

❓ Explaining

5. Why did the colonists enslave native peoples?

_____

_____

_____

_____

Brazil

Lesson 2: History of Brazil, *continued*

⚙ Sequencing

6. Why did the Portuguese import Africans to Brazil?

_____

_____

_____

☑ Reading Progress Check

7. Why did King John III send Jesuits to Brazil?

_____

_____

_____

⚙ Determining Cause and Effect

8. What event caused Brazil to change from a colony to a kingdom?

_____

_____

_____

🔤 Defining

9. Use the word *emancipate* in a sentence. By what year were all enslaved people emancipated?

_____

_____

_____

A new governor brought more colonists to Brazil, including Jesuit Catholic priests. King John wanted the priests to help the native peoples and to convert them to Christianity. Once converted, the natives were protected from slavery. Jesuits went into Brazil's interior to find more native peoples to convert and protect from slavery. Slave hunters, cattlemen, and prospectors soon followed, spreading development of the land.

In the 1600s, large plantations grew sugarcane, mainly in the northeast. This pushed ranching westward. As other resources were discovered, new colonists arrived. Plantations boomed and development of the interior expanded. A number of important products and natural resources helped Brazil grow.

Year	Product/Resource
1600s	Sugarcane, coffee, cotton
1690s	Gold
1720s	Diamonds

Plantation agriculture and mining required large numbers of workers. When there were not enough native peoples to keep up with the work, the Portuguese began to import Africans. By 1820 there were 1.1 million enslaved people in Brazil, nearly one-third of the total population.

Independent Brazil

Guiding Question *How did Brazil gain independence and become a democracy?*

In 1807, Napoleon invaded Portugal. The Portuguese ruler Dom João, the royal family, and other government leaders fled to Brazil. Rio de Janeiro became the capital, and Brazil changed from a colony into a kingdom. After Napoleon was defeated, Dom João's son Pedro stayed in Brazil as ruler.

In 1822, Pedro declared independence for Brazil and crowned himself Emperor Pedro I. Brazil became a constitutional monarchy, a type of government with a king, queen, or an emperor as head of state. After years of harsh rule, Pedro I was forced from his throne. In 1840 his son Emperor Pedro II began his reign, which would last for 50 years.

In 1850, Brazil stopped importing enslaved people from Africa. In the 1860s, a movement began to **emancipate**, or free, enslaved people in Brazil. By 1888, they all were free.

Brazil

Lesson 2: History of Brazil, *continued*

Brazil's plantation owners did not like losing their enslaved workers. In 1889, they supported the army in overthrowing Pedro II. A new constitutional government was established. Brazil became a republic in which the head of state was elected. In this republic, the right to vote was limited to wealthy property owners.

These wealthy voters came from the southeastern states. They elected the governors who controlled the presidential election. In turn, presidents followed economic policies that benefitted southeastern Brazil where coffee was a major export. Over time, some people became unhappy with how the government favored coffee growers over other rich Brazilians.

In 1930, Getúlio Vargas overthrew the "coffee president" and ruled as a dictator for 15 years. He became a hero by instituting many reforms.

Manufacturing thrived under the presidents after 1950. But in 1964, unrest in the country caused military generals to take power. An elected legislature was allowed, but the army controlled the elections. The military gave up power in 1985.

Today Brazil is a democratic republic. Voting is **compulsory**, which means citizens are required to vote. Because Brazil has a high number of well-supported political parties, **coalition governments** are common. A coalition government is one in which several political parties cooperate to do the work of a government.

Making Inferences

10. Why do you think some presidents were nicknamed "coffee presidents"?

_____

_____

_____

Reading Progress Check

11. Why did Brazil's monarchy come to an end?

_____

_____

_____

_____

Writing

Check for Understanding

1. Informative/Explanatory How were the Nambicuara peoples similar to and different from the other main indigenous peoples of early Brazil?

_____

_____

2. Informative/Explanatory Take the role of a Brazilian living in 1930. Write a paragraph supporting or opposing the takeover of the government by Getúlio Vargas.

_____

_____

_____

_____

networks

Brazil

Lesson 3: Life in Brazil

ESSENTIAL QUESTION

What makes a culture unique?

Terms to Know
hinterland remote inland regions, far from the coasts
metropolitan area a central city and the built-up areas around it
central city the largest or most important city in a metropolitan area
favela makeshift community on the edge of a city

What Do You Know?

In the diagram below, write down what you know about the causes of the destruction of Brazil's Amazon rain forest. Later, after you have read the lesson, add more to the diagram.

```
┌──────────────────┐
│                  │
│                  │──────┐
│                  │      │
└──────────────────┘      │
                          ▼
┌──────────────────┐   ┌──────────────────────────┐
│                  │   │                          │
│                  │──▶│ Destruction of the        │
│                  │   │      Rain Forest          │
└──────────────────┘   │                          │
                       └──────────────────────────┘
┌──────────────────┐      ▲
│                  │      │
│                  │──────┘
│                  │
└──────────────────┘
```

Brazil

Lesson 3: Life in Brazil, *continued*

People and Places

Guiding Question *What cultures are represented by Brazilians?*

Brazil is a mix of several cultures. Nearly 40 percent of Brazil's people have mixed racial ancestry. This is because marriage between people of different races is more acceptable in Brazil than in many other countries. The largest mixed-race group is people with European and African ancestors. The next-largest group is people of European and Native American ancestry.

The smallest mixed-race group is people of African and Native American descent. About 4 million Africans had been enslaved and brought to Brazil by the 1800s. Many escaped into the **hinterland**. This is an often remote region, far from the coasts.

Today about 80 percent of Brazilians live within 200 miles of the Atlantic coast. Brazil's highest population of African and mixed populations lives in the northeast and in coastal cities and towns north of Rio de Janeiro.

Most Brazilians of European descent live in southern Brazil. Native Americans live all over the country. The Amazon rain forest is home to the greatest number, but about half of Brazil's Native Americans live in cities now.

In the 1950s, millions of people migrated from rural areas to cities. They wanted to take jobs in Brazil's growing industries. Today, 89 percent of Brazilians live in or around cities.

Sao Paulo is Brazil's industrial center. Some 17 million people live in its **metropolitan area**, which is the city and the built-up areas around the central city. The **central city** is the largest or most important city in the metropolitan area.

In urban areas, middle-class people live in apartment buildings. Others live in small houses in the suburbs. Wealthy Brazilians live in luxury apartments and mansions. Most of Brazil's large cities also have **favelas**. These are makeshift communities that develop on the edges of cities. Some favelas do not have sewers or running water. In many, there is disease and crime. Sao Paulo and Rio de Janeiro have the most and largest favelas. In Rio, about one-third of the city's residents live in favelas.

| Poor, rural Brazilians with few skills and little education migrate to cities. | → | They cannot afford houses or apartments. | → | They settle on land they do not own and build shacks. |

Marking the Text

1. Highlight the various groups of people in Brazil that have multi-ethnic ancestry.

Contrasting

2. What is the difference between a central city and a metropolitan area?

_____

_____

_____

Summarizing

3. In your own words, describe why favelas developed.

_____

_____

_____

_____

Reading Progress Check

4. Why does Brazil have such a large percentage of people with multi-ethnic ancestry?

_____

_____

_____

Brazil

Lesson 3: Life in Brazil, *continued*

Marking the Text

5. Circle the immigrant group that started farming colonies in southern Brazil. Underline the group that worked on coffee plantations.

Analyzing

6. Why do you think Brazilians are better off in cities than in rural areas?

_____

_____

_____

_____

_____

Reading Progress Check

7. Describe one element of Brazil's culture. Explain why that element of culture is important to Brazilians.

_____

_____

_____

_____

_____

_____

_____

_____

People and Cultures

Guiding Question *What is it like to live in Brazil?*

Brazilians have a reputation for accepting other people's differences. Personal warmth, good nature, and "getting along" are valued in Brazilian culture.

Until the late 1800s, nearly all European immigrants in Brazil were from Portugal. Then other immigrants arrived. These new residents worked in many different areas in their new land.

Italians	Worked on coffee plantations
Germans	Started farming colonies in southern Brazil
Japanese	Worked in agriculture in the Brazilian Highlands
Middle Easterners	Became involved in commerce in cities and towns

The mix of people has given Brazil a unique culture. Portuguese is still Brazil's official language, but it is different from the language spoken in Portugal. Thousands of words and expressions have come from ethnic groups and indigenous peoples.

About two-thirds of Brazilians are Roman Catholic. Most of the rest of the population follow the Protestant faith. Islam and Eastern religions are growing in numbers.

There are many African influences on Brazilian culture. These include foods, popular music, and dance, especially the samba. Brazilians blended samba with jazz to create a type of music called the bossa nova.

Each February, Brazilians celebrate a four-day holiday called Carnival. In Rio, costumed Brazilians ride floats in parades, accompanied by lively music and samba dancers.

Most rural families are poor, but they retain close family ties. They work on plantations or small farms and live in one- or two-room houses made of stone or adobe. Adobe is a clay brick dried and hardened in the sun. They eat beans, cassava, rice, and stew.

Many city dwellers are poor too, and they eat a similar diet. In general, though, people in the industrial cities of southern Brazil have a better life than people in the more rural northeast. Many city workers have good jobs and enjoy a decent quality of life. Soccer is Brazil's most popular sport and is played almost everywhere on a daily basis.

networks

Brazil

Lesson 3: Life in Brazil, *continued*

Contemporary Brazil

Guiding Question *What challenges does Brazil face?*

Brazil has the world's seventh-largest economy. It ranks among the leaders in mining, manufacturing, and agriculture. This produces great wealth for some people and a growing middle class. However, 1 in 10 Brazilians must live on less than $2 a day.

Education is another challenge. School is free up to age 17, yet 60 percent of Brazilians have only four years of schooling or less. Brazil's government is trying to improve education at all levels.

The government wants to colonize the country's sparsely populated interior. It also wants to relieve poverty and overcrowding. To do this, several highways have been built across the country. Also, poor rural Brazilians have been offered free land in the Amazon if they will develop it.

Development is a great concern for the future of the rain forest, though. Logging has been a problem for years, but new roads have increased the destruction. About 15 percent of the rainforest is already gone, and the rate of its destruction has attracted worldwide attention. In addition, the soil is often poor and cannot support crops.

Energy is not a concern, however. Large power plants on major rivers produce most of Brazil's energy. In the 1970s, the high cost of imported oil caused the government to substitute ethanol for gasoline. Recent discoveries of oil and natural gas off Brazil's coast give the country all the energy it needs.

? Explaining

8. Why is energy not a concern for Brazil?

_____

_____

_____

✓ Reading Progress Check

9. What are reasons for allowing development in the Amazon rain forest?

_____

_____

_____

_____

_____

Writing

Check for Understanding

1. Informative/Explanatory How has Brazil's African heritage affected its culture today?

_____

_____

2. Narrative Write a paragraph comparing and contrasting living in the country with living in the city in Brazil.

_____

_____

_____

networks

The Tropical North

Lesson 1: Physical Geography of the Region

ESSENTIAL QUESTION

How does geography influence the way people live?

Terms to Know
elevation the measurement of how much above or below sea level a place is
trade winds steady winds that blow from higher latitudes toward the Equator
cash crop a farm product grown for export

Where in the World: The Tropical North

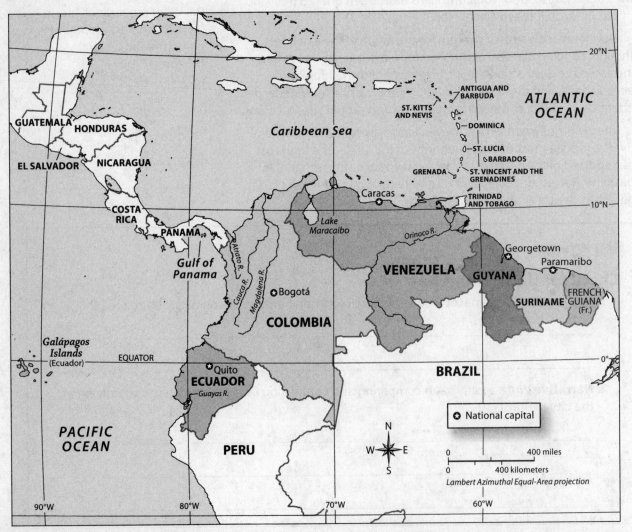

The Tropical North

Lesson 1: Physical Geography of the Region, *continued*

Landforms and Waterways

Guiding Question *What are the major physical features of the Tropical North?*

South America's Tropical North consists of five countries: Ecuador, Colombia, Venezuela, Guyana, Suriname, and French Guiana. The first three have the region's most diverse physical geography. The Andes mountain ranges run through each of them. Some peaks have **elevations** more than 18,000 feet (5,486 m). Elevation is the height above sea level. Cotopaxi in Ecuador is the world's highest active volcano. The Sierra Nevada de Santa Marta in Colombia are the world's highest coastal range.

Colombia has coastlines on both the Pacific Ocean and the Caribbean Sea. Its mountains make travel between the coasts difficult. So does the Darién, a wilderness region of deep ravines, swamps, and dense rain forest.

West of the Andes, Colombia and Ecuador have narrow coastal lowlands. East of the mountains, more lowlands extend into Peru, Brazil, and Venezuela. The southern half is part of the Amazon Basin. The northern half is a grassy plain called the Llanos. This plain also covers most of northern Venezuela.

The Guiana Highlands are a heavily forested region of hills, low mountains, and plateaus. They stretch from Southern Venezuela east into Guyana, Suriname, and French Guiana. Rain forest covers most of this region. There is a narrow band of low and sometimes swampy plains along the Atlantic coast.

The Orinoco River in Venezuela is the continent's third-longest river. Colombia's two main rivers are the Magdalena and the Cauca. They are important routes into the country's agricultural and industrial interior. Commercial ships can navigate most of the length of these rivers. The Guayas River has made Guayaquil Ecuador's largest city and a major seaport.

The Galápagos Islands lie 600 miles (966 km) west of Ecuador. These islands belong to Ecuador. They are home to many unusual animals because the island is so isolated. Today they are protected.

Climates

Guiding Question *How and why do climates vary in the Tropical North?*

South America's Tropical North lies along the Equator. There are a variety of climates due to differences in elevation or location. Others are caused by ocean currents or winds.

Defining

1. What is the definition of *elevation*?

_____

_____

Explaining

2. Why is travel between Colombia's two coasts difficult?

_____

_____

_____

_____

Marking the Text

3. Read the text on the left. Highlight the names of the countries where the Guiana Highlands are located.

Reading Progress Check

4. How do Colombia's rivers help the nation's economy?

_____

_____

_____

_____

_____

networks

The Tropical North

Lesson 1: Physical Geography of the Region, *continued*

 Marking the Text

5. Read the text on the right. Underline the sentences that describe the climate of the Chocó region of Colombia.

✓ **Reading Progress Check**

6. How do the climates of the Pacific coast, the Atlantic coast, and the Caribbean coast differ?

_____

_____

_____

_____

_____

 Drawing Conclusions

7. Why do you think the two nations that are most diverse in natural resources are also the richest countries in the Tropical North?

_____

_____

_____

_____

The region's coasts, interior lowlands, plains, and highlands all have some type of tropical climate. This means that they are warm throughout the year.

Tropical Climates	
tropical monsoon	short dry season and long wet season with heavy rainfall
tropical wet-dry	high annual rainfall; most rain between May and October
tropical rain forest	no dry season

There are coastlines on the Pacific Ocean, the Caribbean Sea, and the Atlantic Ocean. Much of the coastal and eastern lowlands of Ecuador and Colombia have a tropical monsoon climate. In the Chocó region of Colombia, it rains more than 300 days a year. More than 400 inches (1,016 cm) of rain fall each year. It is one of the wettest places on Earth.

The Llanos of Colombia and Venezuela have a tropical wet-dry climate. They get 40 to 70 inches (102–178 cm) of rain a year. The Guiana Highlands have a tropical monsoon climate in some places. In others, there is a tropical rain forest climate.

Guyana, Suriname, and French Guiana have the same climate as Venezuela's highlands. Their coasts are cooled by the **trade winds**. These are steady winds that blow from higher latitudes toward the Equator. The cooler climate of the Caribbean coast of Venezuela and Colombia gets less than 20 inches (30 cm) of rain a year.

Mountain climates depend on the elevation. From 3,000 to 6,500 feet (914 to 1918 m), there is moderate rainfall. Temperatures average 65° to 75°F (18° to 24°C). Above this, it is colder. Above about 10,000 feet (3,048 m), average daily temperatures are below 50°F (10°C). Wind, fog, and light drizzle are common. Vegetation is mainly grasses and hardy shrubs. Above 15,000 feet (4,572 m), the ground is covered with snow and ice year round.

Natural Resources

Guiding Question *Which natural resources are most important to the economies of the Tropical North's countries?*

Tropical rain forests cover much of the North. However, the region's physical geography and lack of roads make it difficult to use this resource. The North's largest countries, Venezuela and Colombia, are also its richest and most diverse in natural resources, as well.

The Tropical North

Lesson 1: Physical Geography of the Region, continued

Oil is found across much of the Tropical North. Venezuela is South America's top producer of oil. It also has large gas deposits and is South America's second-largest coal producer. Columbia is South America's largest coal producer. and the third-largest oil producer. Oil makes up 40 percent of Ecuador's exports.

Gold and diamonds are found throughout the region, with large amounts in some areas, as well as many other minerals and gems.

Mineral and Gem Production	
Gold	Colombia, Ecuador, Venezuela,
Diamonds	Colombia, Venezuela, Guyana, Suriname
Emeralds	Colombia
Bauxite	Guyana, Venezuela, Suriname
Copper, iron ore, other minerals	Colombia, Venezuela, Guyana, Suriname

Ecuador and Colombia grow bananas and coffee as their main **cash crop**, or farm product grown for export, as well as corn, potatoes, beans, and cassava. Colombia produces rice, wheat, sugarcane, cotton, and cattle for sale. Venezuela grows coffee for export, as well as corn and rice, and some ranching is done there. Only about 10 percent of Venezuelans are farmers.

In Guyana, Suriname, and French Guiana, most of the land is covered by rain forest. There is little farming done there and little in the way of oil or mineral resources.

 Marking the Text

8. Highlight the three countries in the region that produce the most oil.

 Marking the Text

9. Highlight the *cash crops* that are important to the economy of the Tropical North.

✓ **Reading Progress Check**

10. Which fossil fuel, mineral, and gem are most widespread in the Tropical North?

_____

_____

_____

Writing

Check for Understanding

1. **Informative/Explanatory** Choose a major geographical feature of the Tropical North and write three facts about that feature.

_____

_____

2. **Informative/Explanatory** How are climate and elevation related in the Tropical North?

_____

_____

The Tropical North

Lesson 2: History of the Countries

ESSENTIAL QUESTION
Why does conflict develop?

Terms to Know
immunity the ability to resist infection by a particular disease
encomienda the Spanish system that allowed Spanish colonists to demand labor from the Native Americans who lived in a certain area
hacienda a large estate

When did it happen?

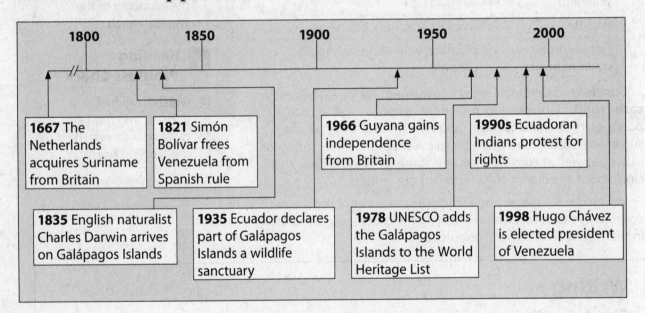

1667 The Netherlands acquires Suriname from Britain

1821 Simón Bolívar frees Venezuela from Spanish rule

1966 Guyana gains independence from Britain

1990s Ecuadoran Indians protest for rights

1835 English naturalist Charles Darwin arrives on Galápagos Islands

1935 Ecuador declares part of Galápagos Islands a wildlife sanctuary

1978 UNESCO adds the Galápagos Islands to the World Heritage List

1998 Hugo Chávez is elected president of Venezuela

Marking the Text

1. Read the text on the right. Highlight the names of Native American groups that lived in the Tropical North.

Early History and Colonization

Guiding Question *How did Europeans colonize the Tropical North?*

Indigenous peoples have lived in this region for thousands of years. Along the Caribbean and Atlantic coasts were Carib, Arawak, and other hunter-gatherer peoples. The Cara and other peoples built fishing villages along the Pacific coast. Groups like the Chibcha and Quitu settled in mountain valleys in the Andes. They farmed, made cotton cloth and gold ornaments, and traded with the Inca. In the late 1400s, some of these groups were conquered and became part of the Inca Empire.

The Tropical North

Lesson 2: History of the Countries, *continued*

Spanish adventurers landed on the Caribbean and Atlantic coasts in the early 1500s. When they found no gold there, they lost interest. Spain first settled in Venezuela in 1523 and Colombia in 1525. They invaded Ecuador in 1534, after conquering the Inca. By the mid-1500s, Spain controlled all of what is now Ecuador, Colombia, and Venezuela. They did not colonize east of Venezuela.

The Spanish set up governments in their new colonies. Their capitals were located where there were already Native American settlements. Most were inland, at higher elevations. Climates were milder than on the tropical coasts. In the 1700s, the Spanish placed Venezuela, Ecuador, and Colombia under a single government.

Native Americans suffered greatly under Spanish rule. Thousands died of European diseases. They had no natural **immunity**, or protection against these illnesses. Other Native Americans were forced to work for Spanish colonists under a system called the **encomienda**. They often worked in mines and for large estates called **haciendas**. Most haciendas grew coffee, tobacco, sugarcane, or other cash crops. Some were cattle ranches.

The British, Dutch, and French fought over and colonized Guyana, Suriname, and French Guiana. The British and Dutch started sugar plantations. They also brought the first enslaved Africans to the area. Control of these colonies changed hands and names several times in the 1600s and 1700s.

Great Britain	British Guiana → Guyana
Netherlands	Dutch Guiana → Suriname
France	French Guiana

Independence

Guiding Question *How did Spain's colonies become independent countries?*

By the late 1700s, many Spanish colonists wanted independence. They got their chance when the French conquered Spain in 1808. Ecuadorans rose up in 1809. Columbians and Venezuelans soon followed. At first, the war was between groups that wanted to be independent and groups that stayed loyal to Spain. Spain expelled the French in 1814. It then tried to restore order in South America.

The colonies resisted Spain's efforts to reestablish control. Venezuelan Simón Bolívar led the revolt in the north. In 1819, Bolivar united Venezuela, Colombia, Panama, and Ecuador. He became the first president of the new republic of Gran Colombia.

? Explaining

2. Why did the Spanish locate most of their capitals inland?

_____

_____

_____

A♭c Defining

3. How were the *encomienda* and the *hacienda* related?

_____

_____

_____

✓ Reading Progress Check

4. Which European nations founded colonies in the Tropical North, and which countries did each colonize?

_____

_____

_____

✐ Marking the Text

5. Read the text on the left. Highlight the names of the countries that united to form Gran Colombia.

networks

The Tropical North

Lesson 2: History of the Countries, *continued*

Copyright © The McGraw-Hill Companies, Inc.

⚙️ **Sequencing**

6. Using the map on this page, create a time line showing the dates of independence for each country in the region.

_____

_____

_____

_____

_____

✏️ **Marking the Text**

7. Using the map on this page, circle the name of the country that is not independent.

☑️ **Reading Progress Check**

8. How did British, Dutch, and French colonists find workers after slavery ended in their colonies?

_____

_____

_____

_____

_____

_____

_____

_____

After Bolívar's death in 1830, Ecuador and Venezuela became independent countries. In the early 1900s, Panama became independent from Colombia.

Independence did not bring democracy and peace. Wealthy landholders and wealthy city businesspeople competed for control of the government. Conflict over the role of the Catholic Church in society added to unrest. There were civil wars in Colombia and Venezuela. Dictators have often ruled Ecuador, Colombia, and Venezuela.

British, Dutch, and French Guiana remained colonies. In 1838 the British abolished slavery in their colony. The French and the Dutch followed in 1863. British and Dutch plantation owners then recruited workers from India and China. The Dutch also brought workers from their colony in Indonesia. The immigrants worked on sugar, rice, coffee, or cacao plantations. After a required length of time, they were free.

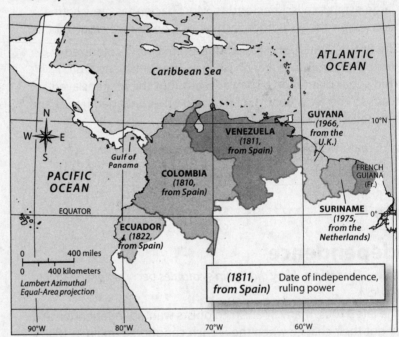

Challenges and Change

Guiding Question *What challenges do the countries of the Tropical North face?*

Independence did not bring an end to trouble. Political and social problems continued to plague Ecuador, Colombia, and Venezuela through most of the twentieth century. Venezuela, for example, did not have a peaceful transfer of power until 1969.

The Tropical North

Lesson 2: History of the Countries, *continued*

The British granted self-government to their colony of Guyana in 1891. In 1953 colonists were given the right to vote and to elect a legislature. Guyana gained independence in 1966. Colonists in Dutch Guiana received the right to vote in 1948. They gained self-government in 1953. The colony became the independent country of Suriname in 1975. The people of French Guiana became French citizens and gained the right to vote in 1848. Since 1945, the colony has been an overseas department, or district, of France.

The lack of strong, stable governments has led to major unrest. Violence between feuding political groups took as many as 200,000 lives in Colombia between 1946 and 1964. Ecuador has not been able to control its remote region in the Amazon Basin. Colombian rebels have built camps in Ecuador and Venezuela. This has led to tension between Colombia and its neighbors.

A border dispute between Guyana and Venezuela was finally settled in 2007. Another border dispute broke out after Suriname's independence. Guyana's African and South Asian populations have competed for power. This has caused years of unrest.

Suriname has also faced internal unrest. The military took power from civilian leaders in 1980 and again in 1990. Rebel groups of Maroons tried to overthrow the government. Maroons are descendants of escaped slaves. In response, the army killed thousands of Maroon civilians. Thousands fled to French Guiana.

Marking the Text

9. Read the text on the left. Underline the sentence that explains the current political status of French Guiana.

Reading Progress Check

10. Which of the region's nations have experienced serious internal unrest since gaining independence?

_____

_____

_____

_____

Writing

Check for Understanding

1. Informative/Explanatory How did conflicts in society lead to independence for Spain's colonies and cause unrest afterward?

_____

_____

_____

2. Informative/Explanatory Why do the Tropical North's nations have a history of tense relations and internal unrest?

_____

_____

_____

networks

The Tropical North

Lesson 3: Life in the Tropical North

ESSENTIAL QUESTION
What makes a culture unique?

Terms to Know
Creole a group of languages developed by enslaved people that is a mixture of French, Spanish, and African
tariff tax on imported goods

What Do You Know?

In the first column, answer the questions based on what you know before you study. After this lesson, complete the last column.

Now...		Later...
	What ethnic groups live in the Tropical North?	
	Where do most of the people of the Tropical North live?	
	What is the Tropical North's culture like?	
	What are some of the challenges that the countries of the Tropical North face?	

🔤 Defining

1. What is the definition of *mestizo*?

_____

_____

People and Places

Guiding Question *What ethnic groups populate the Tropical North, and where do they live?*

Ecuador has the largest indigenous population of any country in the Tropical North. One quarter of the population is Native American. Mestizos are a mixture of white and Native American. Counting mestizos, 90 percent of Ecuadorans have some Native American ancestry.

Lesson 3: Life in the Tropical North, *continued*

In Venezuela and Colombia, 20 percent of the populations are white. Only 1 to 2 percent are Native American. Two-thirds of Colombians and Venezuelans are mestizos. The African populations of Venezuela, Ecuador, and Colombia are small. However, 15 percent of Columbians have mixed African and European ancestry.

Descendants of laborers from India are Suriname's largest group. They make up nearly 40 percent of the population. An equal number of people are of African and mixed-African descent. There is a large Indonesian population. Whites and Native Americans total less than 5 percent of the population.

Native Americans make up almost 10 percent of Guyana's population. A third of the population is African. East Indians account for more than 40 percent. There is no significant white population. About one out of six Guyanese is of mixed ancestry.

People of mixed race make up most of French Guiana's population. There are small groups of French, Native Americans, Chinese, East Indians, Laotians, Vietnamese, Lebanese, Haitians, and Africans.

Most people in the Tropical North live in cities. Bogotá, Colombia, is home to almost 5 million people. About 20 percent of Colombia's people live in the Caribbean lowlands. The Pacific coast is sparsely settled.

Most of Ecuador's Native Americans live in or around Quito. Most other Ecuadorans live along the coast. Venezuelans began moving to cities in the mid-1900s. More than 90 percent of the country's people live in Caracas and other cities along the coast.

Guyana, Suriname, and French Guiana are sparsely populated. Nearly everyone lives along the coast. Suriname's capital is home to more than half of the country's population. Most Guyanese live in small farm towns on the coast.

People and Cultures

Guiding Question *What is the Tropical North's culture like?*

Languages in Guyana, Suriname, and French Guiana reflect their colonial heritage and ethnic populations. **Creole** is widely spoken. It is a group of languages that enslaved Africans developed to help them communicate. Most people in Guyana speak English. In Suriname, the official language is Dutch. However, it is spoken only as a second language. Native American languages, Hindi, and other South Asian languages are heard in both countries.

 Analyzing

2. Why is the Tropical North home to so many ethnic groups?

_____
_____
_____
_____
_____
_____

Marking the Text

3. Read the text on the left. Highlight the names of major cities in the Tropical North.

 Reading Progress Check

4. Where do the greatest number of people in the Tropical North live?

_____
_____
_____
_____
_____

Marking the Text

5. Read the text on the left. Highlight the definition of the word *Creole*.

networks

The Tropical North

Lesson 3: Life in the Tropical North, continued

 Marking the Text

6. In the text, highlight the names of languages that are spoken in the countries of the Tropical North.

? Describing

7. How do Native American cultures continue to influence the Tropical North?

_____

_____

_____

☑ Reading Progress Check

8. What language and religion are most common in the Tropical North?

_____

_____

A♭c Defining

9. How would ending *tariffs* between UNASUR's member nations help their economies?

_____

_____

_____

Spanish is the official language of Ecuador, Colombia, and Venezuela. Native American languages have created regional differences in Ecuadoran Spanish. Colombians, however, have preserved the purity of the Spanish language.

The religions of the Tropical North are equally diverse. Native Americans practice indigenous religions in all countries. However, most people practice religions reflecting the ethnic variety and colonial heritage of the region.

Ecuador	90% Roman Catholic
Venezuela	90% Roman Catholic
Colombia	90% Roman Catholic
Suriname	equal numbers Roman Catholic, Protestant, Hindu, Muslim
Guyana	Protestant, Hindu, with Catholic and Muslim minorities

Each country's foods, music, and other cultural elements reflect its ethnic and religious makeup. Venezuela, Colombia, and Ecuador celebrate Carnival. However, the festivities are not as colorful as those of Brazil. Many Andes communities celebrate regional religious festivals.

Culture often differs by geographic area. Native Americans in mountain regions weave baskets and cloth. They play Andean music using traditional instruments. A dance called the *cumbia* blends Spanish and African heritage. It is popular along the coast of Colombia and Venezuela. Other Venezuelan coastal music shows Caribbean island influences. Maracas and guitars make the music of the Llano.

Ongoing Issues

Guiding Question *What challenges do the countries of the Tropical North face?*

Many people in the Tropical North are poor. The region's natural resources have mostly benefited a wealthy few. This has led to tensions within and between countries.

Many South American leaders believe that trade will strengthen their countries' economies. In 2008 the countries of South America joined to form the Union of South American Nations (UNASUR). UNASUR has several goals, including ending **tariffs**, or taxes on imported goods, between member nations. It also would like to see the region adopt a single currency similar to the euro.

The Tropical North

Lesson 3: Life in the Tropical North, *continued*

Another challenge is improving the region's relationship with the United States. The relationship has been rocky in the past. The United States and the Colombian government are working together to stop the flow of illegal drugs.

In 1998 Venezuelans elected Hugo Chávez as president. He often criticized the United States. He became friendly with anti-U.S. governments in Cuba and Iran. He promised to use Venezuela's oil income to improve conditions for the country's poor. In 2009 he seized control of U.S. oil companies operating in Venezuela. His rule split Venezuela into opposing groups. Working class Venezuelans supported his policies, but middle-class and wealthy Venezuelans opposed him.

The government of Colombia has had a long and bitter struggle with FARC. That is short for the Revolutionary Armed Forces of Colombia. FARC wants to decrease foreign influence in Colombia. It also wants to help the nation's poor farmers.

In Ecuador, indigenous people protested for rights. When the president did not help them, they organized to win rights. They want access to land, basic services, and political representation.

Marking the Text

10. Underline text that describes how the United States is working with a nation of the Tropical North.

Reading Progress Check

11. How did Hugo Chávez raise tensions between Venezuela and the United States?

_____

_____

_____

Writing

Check for Understanding

1. **Informative/Explanatory** Why are there Hindu and Muslim populations living in northern South America?

_____

_____

_____

2. **Informative/Explanatory** How and why is UNASUR likely to affect the economies and people of the Tropical North's countries?

_____

_____

_____

_____

Andes and Midlatitude Countries

Lesson 1: Physical Geography of the Region

ESSENTIAL QUESTION
How does geography influence the way people live?

Terms to Know
cordillera a region of parallel mountain chains
altiplano the high plains
pampas a treeless grassland
estuary an area where river currents and the ocean tide meet
altitude height above sea level

Where in the World: Andes and Midlatitude Countries

Andes and Midlatitude Countries

Lesson 1: Physical Geography of the Region, *continued*

Andes Countries

Guiding Question *What are the physical features of the Andean region?*

Three countries make up most of the region of the Andes mountains: Peru, Bolivia, and Chile. The Andes are the longest continuous group of mountain ranges in the world. They stretch for 4,500 miles (7,242 km). They are also the tallest mountains in the Western Hemisphere. They consist of parallel ranges called **cordilleras**. In Peru and Bolivia, the two main branches of the Andes lie alongside the **altiplano**, or high plain.

These mountain ranges are the result of collisions between tectonic plates. The Andes are part of the Ring of Fire. Because of this, there are earthquakes and volcanic eruptions throughout much of the Andes. The Andes lie 100 to 150 miles (161 to 241 km) inland from the coast. In most places, the land rises steeply from the ocean, with almost no coastal plain.

On the Atlantic side of South America, there are broad plateaus and valleys. This plain is called the **pampas**. It is treeless grassland with fertile soil eroded from the Andes. This area is good for growing wheat and corn and for grazing cattle.

Coastal Peru and Chile and most of southern Argentina are desert. Wind patterns, the cold Peru Current, and high elevations cause low precipitation. There are places in the Atacama Desert in Peru and northern Chile where no rainfall has ever been recorded. The Patagonia desert in Argentina lies in the rain shadow of the Andes.

The Paraná, the Paraguay, and the Uruguay rivers combine to form the second-largest river system in South America. Only the Amazon system is larger. The Paraná-Paraguay-Uruguay system is especially important to Paraguay, which is landlocked. It provides transportation and makes hydroelectric power possible. Along the Paraguay River is the Pantanal, one of the world's largest wetlands. It has a diverse ecosystem of plants and animals.

This river system flows into the Río de la Plata, which empties into the Atlantic Ocean on the border of Argentina and Uruguay. There is a broad **estuary** where the river meets the Atlantic. This is an area where the ocean tide meets a river current.

There are few large lakes in South America. The largest lake in the Andean region is Lake Titicaca, on the altiplano between Bolivia and Peru. It is the highest lake in the world that is large and deep enough to be used by small ships.

Drawing Conclusions

1. On which side of the Andes do you think agriculture would be most productive? Why?

_____

_____

Marking the Text

2. Read the text on the left. Highlight the names of major rivers of the region.

Defining

3. What is the definition of *estuary*?

_____

_____

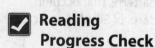

Reading Progress Check

4. How do you think the geography of the Andean region affects the lives of the people who live there?

_____

_____

_____

_____

Andes and Midlatitude Countries

Lesson 1: Physical Geography of the Region, *continued*

Copyright © The McGraw-Hill Companies, Inc.

⚙ **Comparing**

5. How does the climate of the Andes countries compare with that of the midlatitude countries?

_____

_____

_____

🔤 **Defining**

6. Why is *altitude* an important feature in the Andean region?

_____

_____

✏ **Marking the Text**

7. Read the text on the right. Highlight the names of two types of extreme weather patterns that occur in parts of South America.

☑ **Reading Progress Check**

8. Why is the *tierra templada* the most populated climate zone by altitude in the Andean region?

_____

_____

_____

Climate Diversity

Guiding Question *How does climate affect life in the Andean region?*

Climate in the Andes is chiefly determined by **altitude**, or height above sea level. The higher the altitude, the cooler the temperatures. Farming is difficult in the Andean region. Farmers grow crops on hillside terraces. The altitude also makes breathing difficult because the oxygen is thin. The region's inhabitants, as well as native species of plants and animals, have adapted to thinner air.

The midlatitude countries of South America have temperate, or moderate, climates. In Uruguay, rainfall occurs throughout the year. Inland areas, however, are drier than the coast. Argentina is much larger than Uruguay. Its climate varies from subtropical in the north to tundra in the south. Paraguay is generally subtropical. Strong winds often sweep its pampas.

Climate extremes can be experienced in the Andean countries without changing latitude. All that has to change is altitude. The climate changes greatly from lower to higher elevations.

Elevation	Vegetation
tierra caliente ("hot land"), near sea level	bananas, sugarcane, cacao, rice, other tropical plants
tierra templada ("temperate land"), 3,000–6,000 ft. (914–1,829 m)	corn, coffee, cotton, wheat, citrus fruits
tierra fría ("cold land"), 6,000–10,000 ft. (1,829–3,048 m)	forests, grassy areas, potatoes, barley, wheat

Above the *tierra fría* is the *tierra helada*, or "frozen land." Very little vegetation grows in this climate and few people live there.

Every few years, extreme weather occurs in parts of South America. This is caused by changes in wind patterns and ocean currents in the Pacific Ocean. One of these events is called El Niño. It occurs when cold winds from the east are weak, causing the central Pacific Ocean to grow warmer. Water evaporates, and more clouds form. Some areas receive heavy rains. Floods occur in some places, especially along the coast of Peru. Other areas have below-normal rainfall.

In some places, the opposite kind of unusual weather takes place. This event is called La Niña. Winds from the east become strong and cool more of the Pacific.

networks

Andes and Midlatitude Countries

Lesson 1: Physical Geography of the Region, *continued*

Natural Resources

Guiding Question *Which natural resources are important to the region?*

The Andean and midlatitude countries are rich in natural resources. Energy resources are especially important. Bolivia has the second-largest natural gas reserves in South America, as well as large petroleum deposits. In Paraguay, hydroelectric power plants produce most of the country's electricity. The governments want to use these resources to develop and strengthen their economies.

The Andes have one of the world's most important mining industries. Chile leads the world in copper exports. Tin production is important to Bolivia's economy. Both Bolivia and Peru have deposits of silver, lead, and zinc. Peru has gold.

The region's varied geography and climate support a variety of wildlife, including many species of birds and butterflies. The ability of plants and animals to thrive in the region varies with altitude.

A group of mammals called camelids is especially important in this region. Camelids are relatives of camels, but they do not have the typical humps of camels. Two kinds of camelids are the llama and the alpaca. Llamas are the larger of the two. Native Americans throughout the Andes tend herds of llamas. These animals are used to carry goods or pull carts. They are also raised for food, hides, and wool.

Alpacas are found only in certain parts of Peru and Bolivia. The animal's thick, shaggy coat is an important source of wool. Alpaca wool is strong yet soft and repels water. It is used for all kinds of clothing and as insulation in sleeping bags.

Marking the Text

9. Read the text on the left. Highlight the natural resources found in Bolivia.

? Explaining

10. Why is there such a wide variety of wildlife in the Andean and midlatitude countries of South America?

_____

_____

✓ Reading Progress Check

11. What metal is important to Chile's economy?

_____

_____

Writing

Check for Understanding

1. Informative/Explanatory Why do earthquakes and volcanoes occur in the Andes?

_____

2. Informative/Explanatory What effect does altitude have on the climates of the Andean countries of South America?

_____

_____

Andes and Midlatitude Countries

Lesson 2: History of the Region

ESSENTIAL QUESTION
Why do civilizations rise and fall?

Terms to Know
smallpox a highly infectious and often-fatal disease
guerrilla a type of warfare waged by troops who use the local landscape to their advantage
multinational a company that does business in more than one country
coup an illegal seizure of power

When did it happen?

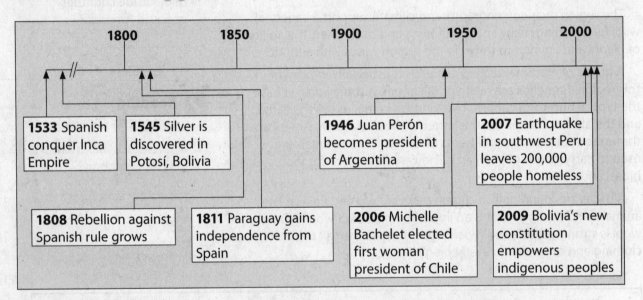

1800	1850	1900	1950	2000

1533 Spanish conquer Inca Empire

1545 Silver is discovered in Potosí, Bolivia

1946 Juan Perón becomes president of Argentina

2007 Earthquake in southwest Peru leaves 200,000 people homeless

1808 Rebellion against Spanish rule grows

1811 Paraguay gains independence from Spain

2006 Michelle Bachelet elected first woman president of Chile

2009 Bolivia's new constitution empowers indigenous peoples

Marking the Text

1. Read the text on the right. Highlight the names of Native American societies that lived in the Andean region before the rise of the Inca.

Early History and Conquest

Guiding Question *How has history influenced the region?*

Before the rise of the Inca in the 1100s, the Andean region was the home of societies such as the Moche, the Mapuche, and the Aymara. These societies were based primarily on agriculture. The Moche settled on the arid coastline of northern Peru. They used irrigation to grow crops such as corn, beans, and other crops.

The Inca first settled in the Cuzco Valley in what is now Peru. By the early 1500s, their empire ran from northern Ecuador southward into Chile. It was home to an estimated 12 million people. This population included dozens of separate cultural groups, who spoke many different languages.

Andes and Midlatitude Countries

Lesson 2: History of the Region, *continued*

The Inca state was called Tawantinsuyu. The name means "the land of four quarters," or provinces. The imperial capital was located where the four provinces of their empire met, at Cuzco.

Inca society was a very structured hierarchy, with the high priest, emperor and army commander at the top. The nobility served the emperor as administrators. Farmers and laborers were at the bottom of society.

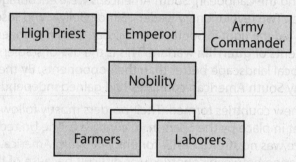

Inca technology and engineering were very advanced. The Inca built extensive irrigation systems, roads, tunnels, and bridges that linked regions of the empire to Cuzco. Today you can still see the remains of Inca cities and fortresses. Some of the most impressive ruins are at Machu Pichu, northwest of Cuzco. They had no written language, but used a counting system called quipu to keep records. A quipu was a series of knotted cords of different colors and lengths. Messengers carrying quipa could travel as far as 150 miles (241 km) a day on the roads. The Inca became very wealthy because of their vast natural resources of gold and silver.

In 1532 a Spanish adventurer named Francisco Pizarro landed in Peru with a small band of soldiers. The soldiers were called *conquistadors,* the Spanish word for "conquerors." He had heard about their fabulous wealth. He had also heard that the Inca empire was badly weakened by a civil war.

Although the Spanish were greatly outnumbered, the Inca were unprepared for Spanish artillery. Within a few years, Spain controlled the entire empire. They took control of the region's precious metals and used the Inca roads to transport goods.

From Peru, the Spaniards branched out to create colonies in Argentina, Chile, and other parts of South America. These colonies became sources of wealth for Spain. Spanish rule of the region continued for nearly 300 years.

The Inca and other Native Americans in the region suffered greatly under the Spanish. Their numbers dropped drastically as a result of **smallpox**, a disease introduced by the Europeans.

Drawing Conclusions

2. Using the diagram on the left, describe the hierarchy of Inca society.

_____

_____

_____

_____

_____

? Identifying

3. What were some of the strengths and achievements of the Inca culture?

_____

_____

_____

✓ Reading Progress Check

4. How were the Spanish under conquistador Francisco Pizarro able to conquer such a mighty empire as the Inca?

_____

_____

_____

_____

Andes and Midlatitude Countries

Lesson 2: History of the Region, *continued*

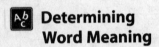

Marking the Text

5. Underline the sentence that tells about two events in North America and Europe that encouraged independence movements in this region.

Determining Word Meaning

6. Use the word *guerrilla* in a sentence.

_____

_____

_____

Reading Progress Check

7. Why are Simón Bolívar and José de San Martín important in the history of the region?

_____

_____

_____

_____

Marking the Text

8. Underline the definition of *multinational*.

Independent Countries

Guiding Question *How did the countries of the Andean region gain their independence?*

By the early 1800s, most of South America had been ruled by Spain for nearly 300 years. The United States threw off British rule, and the French declared a republic. Struggles for independence took place in Mexico and the Caribbean. South Americans were encouraged by these events. Two revolutionary leaders, Simón Bolívar and José de San Martín, led the fight in South America. San Martín pioneered many elements of **guerrilla** warfare. This is the use of soldiers who know the local landscape better than their opponents. By the mid-1800s, many South American countries had gained independence.

Several new countries formed. Their borders mostly followed divisions set in place by the Spanish. In contrast to the United States, there was no strong desire for unity in South America. Communication between countries was difficult because of the mountainous terrain. The rulers of many of the new nations were wealthy aristocrats, powerful landowners, or military dictators. Their attitudes were more European than South American.

The new countries lacked a tradition of self-government. The British colonies in North America had elected representatives in colonial legislatures. The new states of South America did not have this tradition. Nevertheless, they drafted republican constitutions.

Social and economic instability resulted from a huge gap in the distribution of wealth between rich and poor. Bloody conflicts over boundaries or mineral rights led to much loss of life and to weakened economies.

History of the Region in the Modern Era

Guiding Question *What challenges did the countries of the region face in the late 1800s and 1900s?*

Following independence, the Andean and midlatitude countries faced economic and political challenges. Among the economic challenges were developing and controlling resources, building roads and railroads, and creating trade links. The economies of these countries remained tied to countries outside South America.

Beginning in the early 1900s, large U.S. and European **multinational** firms, which do business in several countries, started mining and smelting operations in the region. Wealthy landowners and multinational companies prospered. Many workers and farmers, however, remained poor.

Andes and Midlatitude Countries

Lesson 2: History of the Region, *continued*

Politicians promised they would change the economic system for the better. In 1946 General Juan Perón was elected president of Argentina. His government made economic reforms that benefited workers. However, his government also limited free speech, censored the press, and added to the country's debt.

After overthrowing Perón in 1955, the military government put an end to unrest. They imprisoned thousands of people without trial. Some were tortured or killed, or they simply "disappeared." Argentina was also troubled by conflict with Great Britain over the Falkland Islands, a British overseas territory. In 1982, Argentina lost a brief war and the Falklands remain a British territory today.

In 1970 Chileans elected Salvador Allende as president. He took steps to redistribute wealth and land. The government took over Chile's copper industry and banking system. His reforms were popular with the people, but angered the upper classes. In 1973 he was killed when military officers seized power in a **coup**, an illegal seizure of power. A military dictatorship under General Augusto Pinochet ruled Chile for the next 16 years.

In recent years, democracies have replaced dictatorships. In 2005 Bolivians elected Evo Morales, the country's first indigenous president. He introduced a new constitution and land reform. Chileans and Argentineans have elected women presidents. Both female leaders started efforts to improve human rights and equal opportunity.

Drawing Conclusions

9. What conclusion can be drawn about democratic government from the experiences in Argentina and Chile?

_____

Reading Progress Check

10. After independence, why did the countries in this region continue to experience economic hardship?

_____
_____

Writing

Check for Understanding

1. **Informative/Explanatory** Why did former South American colonies not form a single nation similar to the United States?

_____

_____

2. **Informative/Explanatory** What kinds of challenges faced the Andean and midlatitude nations of South America in the 1800s and 1900s?

_____

_____

_____

Andes and Midlatitude Countries

Lesson 3: Life in the Region

ESSENTIAL QUESTION
What makes a culture unique?

Terms to Know
pueblo jóven a shantytown with poor housing and little or no infrastructure built outside a large metropolitan area

What Do You Know?

In the first column, answer the questions based on what you know before you study. After this lesson, complete the last column.

Now...		Later...
	Where do most people in the Andean and midlatitude countries live?	
	How do three centuries of Spanish rule continue to influence the region?	
	How do Native American customs continue to affect life in the region?	
	What kinds of challenges do the Andean and midlatitude countries face?	

Andes and Midlatitude Countries

Lesson 3: Life in the Region, *continued*

People and Places

Guiding Question *What are the major population patterns in the Andean region?*

Three centuries of colonization by Spain have left their mark on the population of the region. Enslaved Africans were brought as laborers, especially in Peru. Since independence, immigrants from many European countries have settled in South America. Immigrants also came from Asia.

Today, Bolivia and Peru have large Native American populations. Chile and Argentina have many people of European ancestry. They include people of Spanish, Italian, British, and German backgrounds. Peru has many people of European descent. There are also people with origins in Japan and other parts of Asia.

Population patterns reflect the changing impact of politics, economics, availability of natural resources, and jobs. As is the case elsewhere on the continent, the population of the Andean and midlatitude countries is densest in the coastal areas. These areas have fertile land, good climate, and easy transportation. The areas of mountains and tropical rain forests make transportation and communication difficult, discouraging settlement.

Urban and rural areas have great differences in population density. Although Argentina is not densely populated, the area around Buenos Aires is.

Buenos Aires

- Largest city in the region
- Capital of Argentina
- Resembles a European city
- Population: 2.8 million (central city)
                11.5 million (metropolitan area)
- Home to more than one out of four Argentineans

Buenos Aires and many other large cities in the region have shantytowns. These communities often spring up on the outskirts of a city. Poor people migrate from remote regions seeking a better life. They cannot afford housing, so they settle in shantytowns. They build shacks from scraps of sheet metal, wood, and other materials. Shantytowns often lack sewers, running water, and other services. Crime tends to be widespread in them. In Lima, the capital and largest city of Peru, the shantytowns are called **pueblos jóvenes**. The name means "young towns."

Marking the Text

1. Read the text on the left. Underline the sentence that identifies the national origins of the European populations of Chile and Argentina.

? Describing

2. What is life like in the region's shantytowns?

_____

_____

_____

A♭c Defining

3. Why might someone live in a *pueblo jóven*?

_____

_____

_____

✓ Reading Progress Check

4. In what areas of the region is population the densest? Why?

_____

_____

_____

_____

netw🌐rks

Andes and Midlatitude Countries

Lesson 3: Life in the Region, *continued*

Marking the Text

5. Read the text on the right. Highlight Native American cultural influences in the region.

Defining

6. What is magic realism?

_____

_____

_____

_____

Identifying

7. What is the *compadre* relationship?

_____

_____

_____

_____

Reading Progress Check

8. How is Kallawaya medicine different from the modern medicine that is practiced in most Western countries?

_____

_____

_____

_____

People and Cultures

Guiding Question *How do ethnic and religious traditions influence people's lives?*

Although many people trace their ancestry back to Europe, Asia, and Africa, Native American groups still thrive in the region. The Guarani live mainly in Paraguay, but people of Guarani descent can also be found in Argentina, Bolivia, and Brazil. Guarani customs and folk art are an important part of the culture of Paraguay. Guarani is one of the country's official languages. A related language is Sirionó, which is spoken in eastern Bolivia. Quechua, the language of the Inca, is still widely spoken in Peru, along with Spanish.

In Bolivia, the custom of Kallawaya medicine is widespread. Kallawaya healers use traditional herbs and rituals in their cures. In fact, 40 percent of the Bolivian population seeks the help of Kallawayas when they are sick. Either they prefer these natural healers or they cannot afford other doctors.

Under Spanish rule, the Roman Catholic Church was one of the region's most important institutions. The Church's influence continues. Many native peoples combine indigenous rituals and beliefs with Roman Catholicism. In addition, many have adopted Protestant religions.

Traditional arts, crafts, music, and dance thrive in the region. Literature has been especially outstanding. Two Chilean poets have won the Nobel Prize for Literature. Argentinean writers are popular with readers around the world. Many writers use magic realism. This is a style that combines everyday events with magical or mythical elements. It is especially popular in Latin America.

In large cities and towns and in wealthier areas, family life revolves around parents and children. In the countryside, extended families are more common. Throughout the region, the *compadre* relationship is still valued. This is the strong bond between a child's parents and godparents. Social life focuses on family visits, patriotic events, religious feast days, and festivals.

Soccer is the most popular sport in the region, where it is known as football. It is the national sport in many of the countries in the region. The top professional league is made up of teams from 10 South American nations. League teams are eligible for the World Cup and the America Cup. Other popular sports include basketball, golf, boxing, and rugby. Argentina won the first Olympic polo gold medal in 1924. Its polo teams have long dominated international competition.

Andes and Midlatitude Countries

Lesson 3: Life in the Region, *continued*

Ongoing Issues

Guiding Question *How are economic and environmental issues affecting the region?*

It is difficult to build strong economies in the Andean region largely because of the rugged terrain. Many countries rely on agriculture, which is limited and difficult in the mountains. Mining and fishing are also important to the economy of the region. Other activities have also become important. Many people work in factories, producing goods. Others work at jobs in the services industry.

The Andes limit construction of roads and railroads. Yet highways do link large cities. The Pan-American Highway runs from Argentina to Panama and continues northward. Unlike other countries in the region, Argentina has an effective railway system.

In 1991 Argentina, Paraguay, Uruguay, and Brazil signed a trade agreement. In 2011 this agreement merged into the Union of South American Nations. The Union set up an economic and political zone modeled after the European Union. Its goals are to promote free trade and closer political unity.

The Andean and midlatitude countries face many challenges. Air and water pollution is a major problem, especially in the shantytowns. Border disputes have presented challenges for years. Bolivia and Paraguay long fought over a region thought to be rich in oil. A territorial dispute between Peru and Ecuador was marked by armed conflict. Border wars use up resources that could be used to address economic development and environmental concerns.

Marking the Text

9. Read the text on the left. Underline ways in which the Andean and midlatitude countries have begun to cooperate.

☑ Reading Progress Check

10. How does the physical landscape hamper transportation? What actions are being taken to improve transportation?

_____
_____
_____
_____
_____

Writing
Check for Understanding

1. **Informative/Explanatory** What effects did three centuries of Spanish rule leave on the region?

_____

_____

2. **Informative/Explanatory** How have Native American cultures continued to have an influence in the region?

_____

_____

Western Europe

Lesson 1: Physical Geography of Western Europe

ESSENTIAL QUESTION
How does geography influence the way people live?

Terms to Know
dike a large barrier built to keep out water
polder the land reclaimed from building dikes and then draining the water from the land
estuary an area where river currents and the ocean tide meet
Westerlies strong winds that blow from west to east
deciduous trees that shed their leaves in the autumn
coniferous evergreen trees that produce cones to hold seeds and that have needles instead
  of leaves

Where in the World: Western Europe

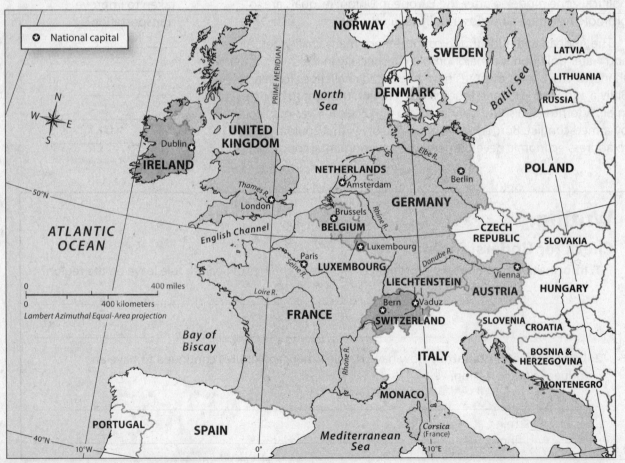

networks

Western Europe

Lesson 1: Physical Geography of Western Europe, *continued*

Landforms and Waterways

Guiding Question *How do the physical features of Western Europe make the region unique?*

Western Europe includes the nations of Ireland, the United Kingdom, France, Germany, the Netherlands, Belgium, Luxembourg, Austria, and Switzerland. It also includes the tiny countries of Monaco and Liechtenstein.

The landscape of the region consists of plains with mountains in some places. Much of Western Europe lies in the Northern European Plain. Massive sheets of ice shaped the plain during the last ice age, which ended about 11,000 years ago. Melting glaciers left behind fertile soil, but also thick layers of sand and gravel. These deposits have eroded into sand dunes along some of the North Sea coastline. The glaciers also left behind areas of poorly drained wetlands along the coasts of the British Isles.

Two mountain ranges separate Western Europe from Southern Europe. They divide the cooler climates of the north from the warm, dry climate of the Mediterranean region to the south.

Range	Pyrenees	Alps
Location	Between France and Spain	France, Switzerland, Austria, Germany
Length	270 mi (435 km)	750 mi (1,207 km)
Tallest Peak	Pico de Aneto 11,169 ft (3,404 m)	Mont Blanc 15,771 ft (4,807 m)

The Pyrenees and the Alps were created by the folding of rocks as a result of plate tectonics. They were also shaped by glaciers. These mountains are younger than other mountains in Europe.

Western Europe has long, irregular coastlines on the Atlantic Ocean and the North, Baltic, and Mediterranean seas. The North Sea is a part of the Atlantic Ocean that separates the island of Britain from the rest of Europe. It is a rich fishing ground for the Netherlands and the United Kingdom. It has long been important for trade. It is also the location of large oil and gas reserves.

The North Sea has helped and hindered the Dutch, the people of the Netherlands. Because 25 percent of the Netherlands is below sea level, the Dutch have built **dikes**, walls to hold back the water. They call the land they reclaim from the sea **polders**. This land is used for farming and settlement. Stormy seas have broken dikes and caused flooding in recent times.

✎ Marking the Text

1. Read the text on the left. Highlight the names of the nations that make up Western Europe.

? Describing

2. How have glaciers shaped the landscape of Western Europe?

_____

_____

_____

_____

⚙ Comparing

3. Compare Western Europe's two major mountain ranges.

_____

_____

_____

_____

? Listing

4. Which bodies of water lie off the coast of Western Europe?

_____

_____

_____

_____

Lesson 1: Physical Geography of Western Europe, *continued*

Marking the Text

5. Read the text on the right. Underline the sentence that tells how Britain is connected to mainland Europe.

✓ Reading Progress Check

6. How did the rivers in Western Europe affect its economic development?

_____

_____

_____

_____

_____

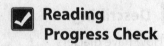

Marking the Text

7. Read the text on the right. Highlight the description of the Mediterranean climate.

✓ Reading Progress Check

8. How do the Westerlies affect the climate in Western Europe?

_____

_____

_____

_____

The British Isles are off the northern coast of France. The English Channel separates southern Britain from northern France. It is a busy sea route connecting the North Sea with the Atlantic Ocean. High-speed trains run through the Chunnel, a tunnel under the English Channel, connecting Britain to mainland Europe.

Western Europe has many rivers and small waterways. Rivers determined the location of important cities, such as London, Paris, and Hamburg. Rivers and canals provide transportation routes for goods and people. Rivers provide water for farming and produce electrical power. They also form political borders. The Thames River in England becomes an **estuary** when it reaches London, then extends to the North Sea. An estuary is where river currents and ocean tides meet. The Rhine is the busiest waterway in Europe. It runs through the most populated region in Europe, from the Swiss Alps to the North Sea.

Climate

Guiding Question *Why is the climate mild in Western Europe?*

Western Europe is located at northern latitudes, but has a milder climate than other places at the same latitudes. This is because most of Western Europe lies in the path of the **Westerlies**. These are strong winds that travel from west to east. They are heated by the warm water of the North Atlantic Current, which originates in the tropical waters of the Caribbean Sea.

This warm, moist air moves inland on the Westerlies. It brings mild temperatures and rain to most of Western Europe throughout the year. Summers are cool, and winters are mild. This climate is known as a marine west coast climate. Because there are no coastal mountain ranges to stop it, the Westerlies blow across the European continent.

Other areas of the region, such as southern France, have a drier climate. Summers are hot and dry, and winters are mild or cool. Most of the rainfall occurs in spring and autumn. This is called a Mediterranean climate.

Natural Resources

Guiding Question *How do the people of Western Europe use the region's natural resources?*

Deposits of coal are plentiful throughout much of Western Europe. Coal fueled machines invented during the Industrial Revolution of the 1800s. Today, coal is less important than other energy sources.

networks

Western Europe

Lesson 1: Physical Geography of Western Europe, *continued*

In 1959 oil and natural gas were discovered under the North Sea. The United Kingdom, the Netherlands, and Germany produce oil and natural gas from the North Sea. Other countries use their rivers to supply energy. Hydroelectricity supplies more than half of Switzerland's electricity needs.

The Northern European Plain has some of the richest soils in Europe. France is Western Europe's leading agricultural producer. Northern France produces wheat. Orchards and vineyards are common in the central and southern parts of the country. Dairy farming is important for the economy of the Netherlands.

The moderate climate and abundant rainfall in most of Western Europe support a variety of plant and animal life. The British Isles have dense forests, grasslands, scrublands, and wetlands. The natural vegetation is mostly **deciduous** forest, or trees that lose their leaves in the fall. The climate on the mainland of Europe is more diverse than that of the British Isles and supports a wider range of plant life.

The drier climates farther inland, as well as the highlands and mountain ranges, support other kinds of plants. **Coniferous** trees, such as fir and pine trees, have cones and needle-shaped leaves. They keep their foliage during the winter. Above the tree line, grasses and shrubs are the most common plants.

Deer, wild boars, hare, and mice are common. Wildcats, lynx, and foxes roam the forests. There are brown bears in the Pyrenees. The number of large animals has decreased in the British Isles, but the islands have more than 200 kinds of birds.

Marking the Text

9. Read the text on the left. Highlight energy sources that are important in Western Europe today.

Defining

10. How do *deciduous* and *coniferous* trees differ from each other?

_____

_____

Reading Progress Check

11. What effect did coal have on the Industrial Revolution?

_____

_____

_____

Writing

Check for Understanding

1. **Informative/Explanatory** Why is the North Sea important to Western Europe?

_____

_____

2. **Informative/Explanatory** Describe the agriculture of Western Europe.

_____

_____

networks

Western Europe

Lesson 2: History of Western Europe

ESSENTIAL QUESTION
Why do civilizations rise and fall?

Terms to Know

smelting the process of refining ore to create metal

feudalism the political and social system in which kings gave land to nobles in exchange for the nobles' military service

Middle Ages the period in European history from about A.D. 500 to about 1450

pilgrimage a journey to a sacred place

Parliament the national legislature of England (now the United Kingdom), consisting of the House of Lords and the House of Commons

industrialized describing a country in which manufacturing is a primary economic activity

Holocaust the mass killing of 6 million Jews by Germany's Nazi leaders during World War II

When did it happen?

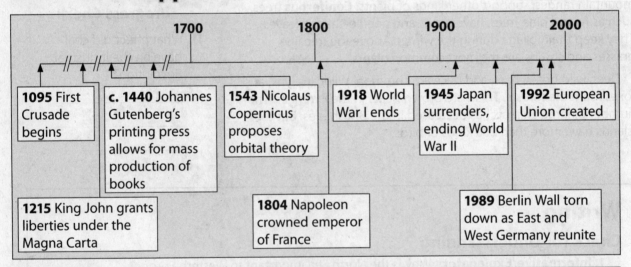

1095 First Crusade begins

c. 1440 Johannes Gutenberg's printing press allows for mass production of books

1543 Nicolaus Copernicus proposes orbital theory

1918 World War I ends

1945 Japan surrenders, ending World War II

1992 European Union created

1215 King John grants liberties under the Magna Carta

1804 Napoleon crowned emperor of France

1989 Berlin Wall torn down as East and West Germany reunite

? Identifying

1. How is bronze made?

   _____

   _____

   _____

History of the Region Through 1800

Guiding Question *How did Western Europe change from a land controlled by loose-knit tribes to a region of monarch-ruled nations?*

Modern humans have lived in Europe for about 40,000 years. Early peoples were hunters and gatherers, then farmers. As populations grew, settlements became towns. People began making metal tools. They discovered how to make bronze by melting and fusing tin and copper, a process called **smelting**.

Lesson 2: History of Western Europe, *continued*

Meanwhile, the Romans spread from Southern Europe into Western Europe. By A.D. 14, they controlled France and most of Germany. They controlled most of Britain within 100 years. The Romans brought their language, Latin, and their technologies.

Over the centuries, Rome's empire in Western Europe weakened. Huns invaded from the east, driving Germanic peoples westward. Rome could no longer protect its lands in Western Europe. Germanic tribes settled there and created kingdoms.

When Christianity became the religion of the Roman Empire in A.D. 312, it spread quickly throughout the empire. The Roman Catholic Church became a major force in Western European life. It helped preserve order and played a key role in education.

As time went on, invaders threatened the region. There were no strong governments. To bring order, the system of **feudalism** arose. Under this system, kings gave land to nobles. The nobles in turn gave kings military service. Today, we call this period of transition between ancient and modern times the **Medieval Age**.

One of the most important rituals in medieval society was the religious **pilgrimage**. This was a visit to places that were central to Christianity, such as Jerusalem. In the late 1000s, Muslims controlled the city. Kings sent great armies to conquer Jerusalem for Christianity. The Crusades, as they were called, failed. Muslims regained control of the city, and their power continued to grow.

The crusaders returned from the east with new ideas. At the same time, the economy of Western Europe was changing. Traders and merchants began to play a bigger role in town life. People with special skills, such as metalsmiths and carpenters, began to organize guilds. Guilds were not as powerful as the noblemen or the Church, but they helped towns grow stronger.

During the Hundred Years' War, which broke out in 1337, France and England fought over land in France. One of the major developments was the spread of a terrible plague, or disease, called the Black Death. The plague raged in Western Europe for four years starting in 1347. Whole towns were wiped out.

The Roman Catholic Church was wealthy and had power over many aspects of society. Many wanted to reform, or change, some Church practices. One such practice was the sale of indulgences, or pardons for people's sins. A German priest named Martin Luther wrote the Ninety-Five Theses, a document attacking this practice. The Church expelled Luther for his beliefs, but his ideas spread. His followers became known as Lutherans. The religious movement he started is called the Protestant Reformation.

 Marking the Text

2. Highlight the countries of Western Europe the Romans controlled.

Abc Defining

3. Describe how *feudalism* worked.

_____

_____

_____

? Describing

4. How did guilds help change life in Western Europe?

_____

_____

_____

_____

? Identifying

5. What was one of the results of the plague?

_____

_____

_____

 Marking the Text

6. Underline the sentences that tell how the Protestant Reformation began.

Lesson 2: History of Western Europe, *continued*

⚙ Finding the Main Idea

7. How did thinking change during the historical period known as the Enlightenment?

_____

_____

_____

⚙ Making Connections

8. How did John Locke's ideas affect the French Revolution?

_____

_____

_____

_____

☑ Reading Progress Check

9. What roles did the Reformation and the Enlightenment play in changing the balance of power in Western Europe?

_____

_____

_____

_____

As the Catholic Church's power weakened, England's kings were being forced to share power with a new lawmaking body called **Parliament**. It was made up of two houses. The House of Lords represented the nobles, and the House of Commons represented common citizens, usually guild members and business owners.

In the 1600s and 1700s a wave of discovery and scientific observation swept over Europe. People began to use reason to observe and describe the world around them. They began to question the authority of kings and the Church. This period is called the Enlightenment.

During this time, explorers were traveling and mapping the world. Astronomers were mapping the solar system. In 1543 Polish astronomer Nicolaus Copernicus suggested that Earth and the other planets revolved around the sun. English philosophers such as John Locke and Thomas Hobbes studied society itself. Locke believed that the best form of government was an agreement between the ruler and the people.

In 1789 France was a powerful country ruled by a king. Most of the people were peasants who lived in poverty. The growing middle class were angry about not having a voice in government. In July of that year, a revolution limited the king's power and ended the privileges of nobles and church leaders. *The Declaration of the Rights of Man and the Citizen* was written. This important document stated that the government's power came from the people, not the king. A few years later, the king was executed.

Not everyone in France supported the revolution. Violence raged for 10 years. In 1799 a young general named Napoleon Bonaparte took military and political control of the country. With a powerful army, he brought much of Europe under French control. In 1814, France's enemies were able to defeat Napoleon and remove him from power.

Change and Conflict

Guiding Question *How did the industrial system change life in Western Europe?*

During the 1800s, some Western European nations **industrialized**. This means they changed from an agricultural society to one based on industry. Cities grew as people moved from the countryside to work in factories. At the same time, some Europeans began to feel strong loyalty to their country. A new, national spirit was rising.

networks

Western Europe

Lesson 2: History of Western Europe, *continued*

A big change took place in Britain between 1760 and 1830. People began to use steam-powered machines for work that had been done by humans or animals. Machines could do the same work at greater speed for lower cost. Machines of the Industrial Revolution also improved farm labor so that fewer people were needed to work the land. People began to leave farms and villages for industrial cities.

Society:	Agricultural	→	Industrial
Population:	Countryside	→	Cities
Power source:	Humans or animals	→	Machines

As nations industrialized, loyalties changed. Former enemies Great Britain and France became closer as Germany gained military strength. As the possibility of war increased, countries formed alliances.

European nations became rivals for colonies and economic power. These rivalries helped lead to World War I. Political changes also contributed as modern nation-states replaced monarchies and empires. World War I began in 1914 and involved all of Europe. It resulted in millions of deaths and great destruction. Germany lost the war and was found guilty of starting it. The winning countries demanded that Germany pay for damages.

The defeat greatly weakened the German economy. Germans believed that they were being punished too harshly for their role in the war. A political radical named Adolf Hitler used the people's anger to gain power. He fueled an aggressive nationalist spirit. By 1933 he was the absolute ruler of Germany.

Hitler and his Nazi Party believed that the Germans were a superior race. They carried out the **Holocaust**, the government-sponsored murder of 6 million Jews. Other minorities also suffered under the Nazis. Hitler and the Nazis wanted to build a new German empire. They began to build a military force to do this.

War came when Hitler's armies began seizing other countries. World War II stretched far beyond Western Europe. Germany allied with Italy and Japan. They were the Axis Powers. The Soviet Union and the United States sided with Britain. They were the Allied Powers. American soldiers fought in Europe, Africa, and the Pacific. A combination of American, British, and Canadian troops invaded France in June 1944 and liberated it from Germany.

✎ Marking the Text

10. Read the text on the left. Underline the sentence that describes a change in loyalties that came about as a result of industrialization.

❓ Identifying

11. Who was Adolf Hitler?

_____

_____

_____

_____

ᴬᵇᵪ Defining

12. What was the *Holocaust*?

_____

_____

_____

_____

❓ Identifying

13. Who were Germany's allies during World War II?

_____

_____

_____

_____

Western Europe

Lesson 2: History of Western Europe, *continued*

Sequencing

14. When and in what area did World War II end first? When and where was its final end?

_____

_____

_____

_____

✓ Reading Progress Check

15. How were the causes of World War I and World War II similar? How were the causes different?

_____

_____

_____

_____

Germany
Italy
Japan

Britain
United States
Soviet Union

With Hitler's death and Germany's surrender in May 1945, the war ended in Europe. However, it continued in East Asia and the Pacific for another three months. It did not end until the United States used atomic bombs on the Japanese cities of Hiroshima and Nagasaki. During the war, between 40 million and 60 million people died. More civilians died than military forces.

Before the war, Britain, France, and Germany were among the most powerful countries in the world. They and other European nations were weakened by the war. After the war, the United States and the Soviet Union emerged as the leading world powers. The United States was a strong ally to nations in Western Europe. The Soviet Union took control of most of Eastern Europe. Germany was split, with Britain, the United States, and France occupying the western part. The Soviet Union controlled eastern Germany. This was the beginning of the Cold War.

For over 40 years, the United States and the Soviet Union were in conflict. War never broke out, but the threat of war always existed. Both sides stockpiled nuclear weapons. In the 1980s, Soviet influence began to weaken. Protest movements spread to European countries under Soviet control. The Cold War ended when the government of the Soviet Union collapsed in 1991.

Writing

Check for Understanding

1. Informative/Explanatory List three things that the Romans brought to Western Europe.

_____

_____

2. Informative/Explanatory How did World War II change the world's power structure?

_____

_____

Western Europe

Lesson 3: Life in Western Europe

ESSENTIAL QUESTION
How do governments change?

Terms to Know
postindustrial describing an economy that is based on providing services rather than manufacturing

---

What Do You Know?

In the first column, answer the questions based on what you know before you study. After this lesson, complete the last column.

Now...		Later...
	How were the Western European nations able to recover following two world wars?	
	How did Western European culture spread around the world?	
	How has the economy of Western Europe evolved since World War II?	

People, Places, and Cultures

Guiding Question *What are the contributions of Western Europe to culture, education, and the arts?*

Political events in the 1900s threatened all of Europe. In order to survive and compete in a changing world, the nations of Western Europe needed to learn to work together. After World War II, they made efforts to do this.

In April 1951, the Treaty of Paris created an international agency to supervise the coal and steel industries. France, West Germany, Belgium, the Netherlands, Luxembourg, and Italy signed this treaty. These six nations created the European Economic Community (the EEC) in 1958 to make trade among the member nations easier.

⇄ Accessing Prior Knowledge

1. What political events in the 1900s made European countries believe they needed to work together?

_____

_____

_____

_____

Western Europe

Lesson 3: Life in Western Europe, *continued*

 Sequencing

2. Make a timeline showing the dates and agencies that led to the formation of the European Union.

_____

_____

_____

_____

✎ Marking the Text

3. Read the text on the right. Highlight the names of the 12 original nations that formed the European Union in 1993.

? Summarizing

4. How do the nations of the European Union work together?

_____

_____

_____

_____

_____

Aᵇc Defining

5. What are the *Indo-European* languages?

_____

_____

In 1967 these countries came together to form the European Commission. In 1971 they were joined by the United Kingdom and Ireland. Denmark, Greece, Spain, and Portugal joined the Commission by the late 1980s.

Those 12 nations formed the European Union, or EU, in 1993. The goal of the EU is to strengthen trade between the countries of Europe. Member nations control their own political and economic decisions. However, they follow EU laws on the use of natural resources and release of pollutants. They also have agreements on law enforcement and security.

The European Union now has 27 members. Eight of those nations are in Western Europe: Austria, Belgium, France, Germany, Ireland, Luxembourg, the Netherlands, and the United Kingdom. East and West Germany reunited when the Soviet Union lost control in the 1980s. Germany is now a strong voice in the EU.

Celts, Saxons, Romans, Vikings, Visigoths, and others fought for control in ancient Western Europe. Those groups faded as the modern nations of Europe began to take shape. Ethnic groups such as the French and Germans now rule entire countries. Their languages are the main languages of those nations. Most of those countries are home to members of other ethnic minority groups. Many of these are immigrants. They often speak the language of their homeland and keep their own culture.

The Indo-European languages are a group of related languages. They are spoken in most of Europe, parts of the world that were colonized by Europeans, and India and other parts of Asia. Two major divisions of Indo-European languages spoken in Western Europe are Romance and Germanic. Romance languages are based on Latin, the language of the Roman Empire.

```
┌─────────────────────────────────────────────┐
│  Indo-European Languages in Western Europe    │
└─────────────────────────────────────────────┘
         ↙                          ↘
┌─────────────────────┐    ┌─────────────────────┐
│  Romance Languages  │    │  Germanic Languages │
└─────────────────────┘    └─────────────────────┘
```

Although English is a Germanic language, about half of its words come from the Romance languages. Not all Western European languages are Indo-European, however. For example, Basque is spoken in the Pyrenees region of France and Spain. It is not related to any other language spoken today. Many Western Europeans speak more than one language—their native language plus English, French, or German.

Western Europe

Lesson 3: Life in Western Europe, *continued*

Christianity continues to be the major religion in Europe. Today, most Western Europeans are either Catholic or Protestant. The Roman Catholic faith is strongest in France, Ireland, and Belgium. Protestant churches are strongest in the United Kingdom and Germany. Many Muslims have immigrated to Western Europe. They follow the religion of Islam. The Holocaust nearly wiped out Europe's Jewish population. Today, Jewish communities are growing in Western Europe, especially in France, the United Kingdom, and Germany.

For centuries, Western Europe has been a world leader in culture and the arts. European explorers spread European culture to other parts of the globe. The arts are an important part of Western European culture. Museums and cultural institutions celebrate each nation's art and history, and national governments support the arts. Western European culture has had a major influence on the rest of the world.

The most important team sport in Western Europe is soccer. Cricket and rugby are popular in the United Kingdom. Mountain climbing, skating, and skiing are popular in the rugged Alps of Switzerland and Austria.

Much of the population of Western Europe lives in cities, so roads are crowded. To relieve traffic congestion and control pollution, much of Europe has turned to high-speed rail lines. The first of these were built in France in 1981. In the 1990s, French rail lines began connecting to other high-speed rail lines. A well-developed highway system also links Europe's major cities. Germany's superhighways, called autobahns, are among Europe's best roads.

Western Europe is one of the wealthiest, most urban, and well-educated regions in the world. Many students go on to attend college. The region contains some of the oldest and most famous universities. Oxford University in England and the University of Paris opened before 1200. Originally, universities were founded to educate the clergy.

Current Challenges

Guiding Question *Why is Western Europe considered a postindustrial region?*

Since the Industrial Revolution, improvements in agriculture have made it possible for fewer people to cultivate larger areas of land. Today, more than half the population of Western Europe lives and works in cities. Even in France, the region's leading agricultural nation, less than 4 percent of the workforce works in agriculture.

✏ Marking the Text

6. Read the text on the left. Highlight the names of major religions that are practiced in Western Europe today

⇄ Activating Prior Knowledge

7. What was the Holocaust?

_____

_____

_____

_____

❓ Identifying

8. What was the original purpose of universities?

_____

_____

_____

✅ Reading Progress Check

9. In what ways do nations of Western Europe support art and culture?

_____

_____

_____

Western Europe

Lesson 3: Life in Western Europe, *continued*

? Explaining

10. What is the source of population growth in Western Europe today?

_____

_____

_____

✓ Reading Progress Check

11. What challenges do the nations of Western Europe face?

_____

_____

_____

_____

_____

In the past few decades, the number of industrial workers has also declined. The industrial, or secondary, sector of the economy employs only about 25 percent of the people. Many more people work in the tertiary sector, or service industries. Service jobs include government, education, health care, financial services, retail, computing, and repair. Every nation in Europe has a **postindustrial** economy. That means that more people work in services than in industry.

In 1900 Great Britain, France, and Germany ruled over empires that extended beyond Europe to Asia, Africa, the Americas, and the Pacific Islands. The two world wars were hard on the region. Then the Cold War kept it on the brink of war for 40 years. Germany, France, and the United Kingdom are still among the seven biggest economies in the world. The European Union helps Western European nations compete with larger economies, such as the United States, China, and Japan. However, the global financial crisis of 2008 had an impact on all of Europe. Governments of the EU disagreed about how to deal with ongoing financial problems.

Most population growth in Western Europe today is from immigration. Germany, France, the Netherlands, and United Kingdom have large Muslim populations. People come from Africa, Asia, and Eastern Europe looking for job opportunities or to escape political oppression. The mix of European and immigrant cultures creates a more diverse culture, but also creates racial and religious tensions.

Writing

Check for Understanding

1. Informative/Explanatory How have the nations of Western Europe adapted to survive the political events of the 1900s?

_____

_____

_____

2. Informative/Explanatory What effects has immigration had on the nations of Western Europe?

_____

_____

_____

networks

Northern and Southern Europe

Lesson 1: Physical Geography of the Regions

ESSENTIAL QUESTION
How do people adapt to their environment?

Terms to Know
glaciation the weathering and erosion caused by the movement of glaciers
fjord a narrow, U-shaped coastal valley with steep sides formed by the action of glaciers
tundra a flat, treeless plain with permanently frozen ground
scrubland area with a dry climate where mostly short grasses and shrubs grow
trawler a large fishing boat

Where in the World: Northern and Southern Europe

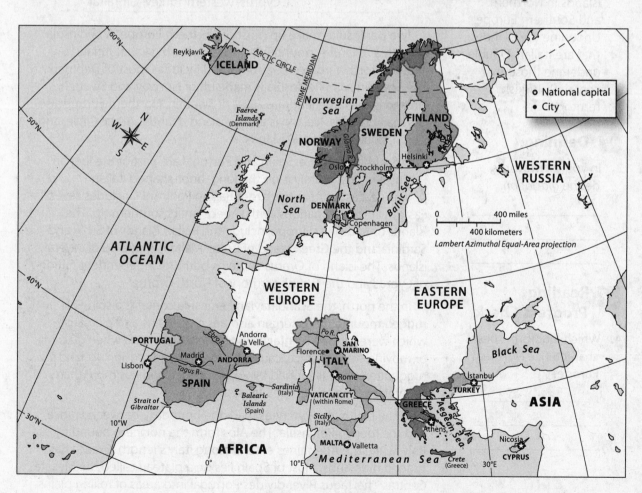

networks

Northern and Southern Europe

Lesson 1: Physical Geography of the Regions, *continued*

Marking the Text

1. Read the text on the right. Underline all of the peninsulas in Northern and Southern Europe.

Marking the Text

2. Read the text on the right. Circle all the islands in Northern and Southern Europe. Underline the islands that are Italian. Double underline those that contain a Turkish territory.

Defining

3. In your own words, define *glaciation*.

_____

_____

_____

✔ Reading Progress Check

4. Which landforms best characterize Northern Europe?

_____

_____

_____

Landforms and Waterways

Guiding Question *How are the landforms in Northern and Southern Europe similar? How are they different?*

Much of Northern Europe is a land of rugged mountains, rocky soils, and jagged coasts. The coast of Southern Europe is dominated by the Mediterranean Sea. The table below lists the countries in both regions.

Region	Countries
Northern Europe	The Nordic Countries: Denmark, Sweden, Norway, Finland, Iceland
Southern Europe	Spain, Portugal, Italy, Greece, Malta, Andorra, San Marino, Vatican City, Cyprus, western Turkey, Gibraltar

Two peninsulas make up most of Northern Europe. A peninsula is an area of land surrounded on three sides by water. Jutland extends northward from Germany and includes most of Denmark. The Scandinavian Peninsula is made up of Norway and Sweden. Finland is the large landmass east of Sweden. Northern Europe also has many islands. Iceland is a large island near the Arctic Circle, and Denmark has about 400 islands.

In Southern Europe, Spain and Portugal are part of the Iberian Peninsula. Most of Italy is on the long, boot-shaped Italian peninsula. East of Italy, the larger Balkan Peninsula includes several Eastern European nations, with Greece on its southern tip. The islands of Southern Europe include the Italian islands of Sicily and Sardinia and the Greek island of Crete. Malta is made up of several islands. The island of Cyprus contains both the independent nation of Cyprus and the Turkish territory of North Cyprus.

In the north, the Scandinavian Peninsula contains a spine of rugged mountains. Its terrain also has mountains and plateaus which were formed by **glaciation**. Glaciation occurs when glaciers, or moving masses of ice, cause weathering and erosion. Iceland's rugged terrain was formed by volcanoes. Iceland has more than 200 volcanoes and Europe's largest glacier.

In the south, the Pyrenees mark the boundary between France and the Iberian Peninsula. The Alps form the northern boundary of Italy, while the Apennines extend along Italy's length. Greece has rugged highlands. Most of Spain lies on a plateau called the Meseta Central. The Tagus River divides Portugal into areas of rolling plains and hills and valleys.

networks

Northern and Southern Europe

Lesson 1: Physical Geography of the Regions, *continued*

Northern Europe has few important rivers, but has a long coastline with many seas and bays. Much of Norway's west coast is dotted by **fjords**, or narrow, water-filled valleys formed by glaciers. The Baltic Sea borders Sweden and Finland.

The most important body of water in Southern Europe is the Mediterranean Sea. It stretches amost 2,500 miles (4,023 km) from the southern coast of Spain to the west coasts of Greece, Turkey, and parts of Southwest Asia. Important rivers in Southern Europe include the Po in Italy, the Ebro in Spain, and the Tagus on the Iberian Peninsula.

Contrasting Climates

Guiding Question *How is the climate of Northern Europe different from the climate of Southern Europe?*

In Northern Europe, the Norwegian Current flows past Norway carrying warm water from the tropics. This gives western Norway a marine climate with mild winters and cool summers. Mountains reduce the air going west, so eastern Norway has colder and snowier winters.

Finland's climate is continental because the seas do not influence it. It has cold winters and hot summers. In northern Finland, the land is mostly **tundra**. This is where subsoil is frozen and only plants such as lichens and mosses can survive.

The Gulf Stream brings warm water to the southern and western coasts of Iceland. This moderates the temperatures in these parts of Iceland. The effect does not reach northern Iceland, though, where temperatures are colder year-round than elsewhere in the country.

Gulf Stream	Brings warm water to southern and western Iceland	Moderates temperatures on coast of Iceland

Most of Southern Europe has a Mediterranean climate. Summers are warm or hot, and winters are cool or mild. Spring and fall are rainy, but summers are dry.

Temperatures are not the same across Southern Europe. Northern Italy is nearer the Alps and has a cooler mountain climate. Winters there are colder than in southern Italy, and snow is heavy at higher elevations. The Meseta Central in Spain has a continental climate. Its high elevation and mountains cause dry winds and drought throughout the year. This brings cold winters and hot summers.

 Marking the Text

5. Read the text on the left. Highlight the vocabulary word that describes an area where only lichens and mosses can survive.

 Marking the Text

6. Underline how the Gulf Stream affects temperatures in Iceland.

? Identifying

7. What kind of climate does most of Southern Europe have?

_____

_____

☑ Reading Progress Check

8. How do landforms and waterways affect the climates of Norway and Italy?

_____

_____

_____

_____

_____

_____

Northern and Southern Europe

Lesson 1: Physical Geography of the Regions, *continued*

? Identifying

9. How do the countries of Northern Europe get much of their energy?

_____

_____

_____

✓ Reading Progress Check

10. How can a thriving fishing industry be a positive and a negative factor for a country?

_____

_____

_____

_____

_____

Natural Resources

Guiding Question *What natural resources are available to the people of Northern and Southern Europe?*

In Southern Europe's dry climate, plants need to be resistant to drought. In fact, many areas are **scrubland**, or places where short grasses and shrubs are the most common plants. Grapes and olives are important crops, while wine and olive oil are important exports.

Northern Europe has forests, so the main plant resource there is wood. Finland exports birch, spruce, pine wood, and paper products. Sweden's forests produce timber, paper, pulp, and furniture. Norway is Europe's biggest exporter of oil and one of its leading suppliers of natural gas. Northern Europe also has rich mineral ore resources, with deposits of iron ore, copper, titanium, lead, nickel, and zinc.

Denmark uses wind turbines to supply much of its electricity. Sweden and Norway use hydroelectric power plants. Because of its volcanoes, Iceland has a great deal of geothermal energy.

Norway has one of the biggest fishing industries in Europe. Today, fewer people work in this industry, because ships that tow large nets behind them are able to catch large amounts of fish. These ships are called **trawlers**.

Spain and Portugal have long Atlantic Ocean coastlines, so fishing is an important industry there. It is also important to countries along the Mediterranean Sea, such as Italy and Greece. However, overfishing and pollution have hurt fish populations in the Mediterranean.

Writing

Check for Understanding

1. **Informative/Explanatory** How are the climates of Northern and Southern Europe alike and different?

_____

_____

_____

_____

_____

networks

Northern and Southern Europe

Lesson 2: History of the Regions

ESSENTIAL QUESTION

Why do civilizations rise and fall?

Terms to Know

city-state an independent political unit that includes a city and the surrounding area

longship a ship with oars and a sail used by the Vikings

pagan a believer in ancient myths with numerous gods

Renaissance the period in Europe that began in Italy in the 1300s and lasted into the 1600s, during which art and learning flourished

---

When did it happen?

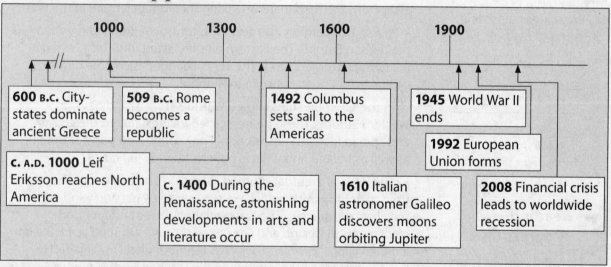

600 B.C. City-states dominate ancient Greece	**509 B.C.** Rome becomes a republic	**1492** Columbus sets sail to the Americas	**1945** World War II ends

c. A.D. 1000 Leif Eriksson reaches North America

c. 1400 During the Renaissance, astonishing developments in arts and literature occur

1610 Italian astronomer Galileo discovers moons orbiting Jupiter

1992 European Union forms

2008 Financial crisis leads to worldwide recession

Early History of the Regions

Guiding Question *Why were early civilizations in Northern and Southern Europe important?*

In ancient Greece, separate communities called **city-states** formed. Each city-state was an independent political unit linked to other city-states by Greek language and culture. Although often jealous of each other, they joined forces to fight a common enemy.

In 490 B.C., the Persian army invaded, and the city-states of Athens and Sparta joined together to defeat the mighty Persian army. After the Persian Wars, Athens emerged as the most developed city-state. It became the first-known democracy, and philosophy and art flourished there.

Abc Defining

1. What is the definition of *city-state*?

_____

_____

_____

_____

Northern and Southern Europe

Lesson 2: History of the Regions, continued

Marking the Text

2. Read the text on the right. Highlight the passage that explains how the Greek city-states were taken over by another kingdom. Circle the name of the king who conquered them.

? Identifying

3. What areas did Rome come to control?

_____
_____
_____
_____
_____

☑ Reading Progress Check

4. How did warfare affect the civilizations of Greece, Rome, and the Vikings?

_____
_____
_____
_____
_____
_____

Years of war eventually weakened the Greek city-states. The king of Macedon took over Greece as well as Asia Minor, Persia, Egypt, and other kingdoms. Alexander the Great, as he was called, created an empire that would last another 300 years.

While Greek city-states were at their height, the city of Rome rose to power and became the Roman Republic. It was run by consuls who were elected to office annually.

By 275 B.C., Rome controlled the Italian peninsula. It then conquered Spain, Sicily, Macedonia, Greece, and Asia Minor. After the military leader Julius Caesar was assassinated, there were a series of emperors in Rome. The Roman Republic became the Roman Empire.

In A.D. 330, the Emperor Constantine moved the capital from Rome to Byzantium in what is now Turkey. The new capital, renamed Constantinople, was closer to trade routes and farther away from the barbarians attacking the Roman Empire in the west.

In A.D. 476, German invaders took control of Rome and ended the Roman Empire. The eastern empire lasted until 1453, when it fell to the Ottoman Turks. The Turks changed the name of the capital from Constantinople to Istanbul.

In the A.D. 700s, ships carried warriors from Scandinavia to Western Europe. These pirates, called Vikings, raided the coasts and spread fear and destruction. They conquered parts of Britain, as well as Ireland and what is now Normandy in France.

The Vikings eventually turned to trading and exploring. They used their **longships**, which were powered by oar and wind, to travel great distances. They crossed the Atlantic and founded settlements in Iceland and Greenland. Around A.D. 1000 Leif Eriksson led the Vikings to Vinland, which is Newfoundland in Canada. He was the first European known to have reached North America.

The Vikings followed a **pagan** religion, which was based on ancient myths and had a number of gods. After about A.D. 1000, Viking groups began converting to Christianity. More Scandinavians stayed home and the threat of Vikings went away. The Scandinavians helped build the kingdoms of Norway, Sweden, and Denmark.

Discovery and "Rebirth"

Guiding Question *How did the Renaissance pave the way for voyages of discovery?*

During the period known as the Middle Ages, many important scientific discoveries of the ancient Greeks were largely forgotten in the west. Many important manuscripts were taken to the east.

Northern and Southern Europe

Lesson 2: History of the Regions, *continued*

When the Byzantine Empire fell in 1453, scholars took many ancient Greek manuscripts west. There the manuscripts were printed on a printing press, which had just been invented. People could now own and read these books. This breakthrough resulted in the **Renaissance**, a period of artistic and intellectual activity. It began in Italy and spread throughout Europe.

In 1609, Italian astronomer Galileo designed a telescope to see the moon and planets. His work helped prove the theory that the planets orbit the sun.

By the 1400s, Europeans wanted to do more business with China and India, but travel by land was long and dangerous. European rulers encouraged sailors and navigators to look for a sea route to Asia. The table below shows some of these explorers and their accomplishments.

Explorer	Country	Year	Accomplishment
Bartholomeu Dias	Portugal	1488	Reached the Cape of Good Hope at the tip of Africa
Vasco de Gama	Portugal	1498	Rounded the Cape and sailed to India
Christopher Columbus	Spain	1492	With three ships, sailed to the Americas

Through its trips to the Americas, Spain became the most powerful country in Europe. Spain built an empire in Mexico, Central America, and South America. In addition, contact between Europe and the Americas led to an exchange of goods called the Columbian Exchange. Europeans brought wheat, coffee, sugar, horses, and sheep to the Americas. They took tomatoes, corn, potatoes, and squash back to Europe.

Meanwhile, Christianity had become identified with Europe, particularly Rome. When the Roman Empire split, so did Christianity. The western branch became the Roman Catholic Church. The eastern branch became the Eastern Orthodox Church.

Other religions spread into Europe, too. In the A.D. 700s, the Moors, an Arabic people, invaded Spain from Northern Africa. They brought the religion of Islam with them. In 1453, the Byzantine Empire fell to the Ottoman Turks, another Muslim people. During the 1520s the ideas of Martin Luther led to the Protestant Reformation. Kingdoms across Northern Europe broke away from the Roman Catholic Church and set up Protestant churches.

Marking the Text

5. Read the text on the left. Underline the definition of *Renaissance*.

Analyzing

6. Why was the printing press such an important invention?

_____

_____

_____

_____

Summarizing

7. Which country was able to create an empire in the Americas? Why?

_____

_____

_____

Reading Progress Check

8. Why did Christianity change as it took hold in Southern and Northern Europe?

_____

_____

_____

_____

Northern and Southern Europe

Lesson 2: History of the Regions, *continued*

Locating

9. How did Spain change during the 1800s? How did Italy change?

_____

_____

_____

Marking the Text

10. Circle the countries that became democracies after World War II.

Reading Progress Check

11. What country did Italy side with during World War II?

_____

History in the Modern Era

Guiding Question *What has been the relationship between Northern and Southern Europe over the last 200 years?*

The 1800s brought sweeping political changes to Northern and Southern Europe, as shown in the table below.

Scandinavian countries	Became prosperous democracies
Spain and Portugal	Lost much of their overseas empires
Greece	Won freedom from the Turks
Italy	United its separate territories, except for San Marino and Vatican City

During the 1900s, Northern and Southern Europe were involved in both world wars. After fighting a civil war in the 1930s, Spain stayed out of World War II. Italy, ruled by dictator Benito Mussolini, sided with Nazi Germany. The Italians were defeated in 1943. The war inEurope ended in 1945.

After World War II, there were many political changes. Italy, Spain, Portugal, and Greece became democracies. Greece also fought a civil war between Communists and anti-Communists.

Since 1945, Scandinavia has had a high standard of living as well as political and social freedoms. Beginning in the 1990s, the nations of Northern and Southern Europe developed closer ties with one another and with other European countries as they joined the European Union.

Writing
Check for Understanding

1. **Informative/Explanatory** How did the invention of the printing press contribute to the Renaissance?

_____

_____

2. **Informative/Explanatory** Why did the Emperor Constantine move the capital of the Roman Empire from Rome to Byzantium?

_____

_____

networks

Northern and Southern Europe

Lesson 3: Life in Northern and Southern Europe

ESSENTIAL QUESTION
How do new ideas change the way people live?

Terms to Know
homogeneous similar
dialect a regional variety of a language with unique features, such as vocabulary, grammar, or pronunciation
welfare capitalism a system in which the government uses tax money to provide services, such as health care and education, to all citizens
recession a period of slow economic growth or decline

What Do You Know?
In the first column, answer the questions based on what you know before you study. After this lesson, complete the last column.

Now...		Later...
	What is the major religion in Southern Europe? What is the major religion in Northern Europe?	
	What kind of economic system is used in Northern Europe?	
	What is the European Union?	

Lesson 3: Life in Northern and Southern Europe, *continued*

? **Explaining**

1. How does a high standard of living affect people in Northern and Southern Europe?

_____

_____

_____

_____

⚙ **Contrasting**

2. What percent of the population of Greece lives in cities? What percent in Spain?

_____

_____

✓ **Reading Progress Check**

3. Why is the population of Northern and Southern Europe growing older?

_____

_____

_____

_____

✏ **Marking the Text**

4. Read the text on the right. Circle the definition of *homogeneous*.

People and Places

Guiding Question *What is the distribution of the populations in Northern Europe and Southern Europe?*

In most places in Northern and Southern Europe, the standard of living has improved. This has helped people live longer, and it has reduced the number of babies who die in their first year of life. The overall birthrate has gone down, though. Why? Europeans have decided to have fewer children. The effect of this is shown in the diagram below.

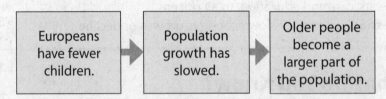

| Europeans have fewer children. | → | Population growth has slowed. | → | Older people become a larger part of the population. |

In Southern Europe, the number of people who live in rural areas has gone down, while the number of people in cities has gone up. In Rome, Milan, and Naples, Italy's three largest cities, the population is about 10 million people total. In Greece, about 60 percent of the people live in cities. In Spain, more than 75 percent live in urban areas. More than 5 million people live in Madrid.

In Northern Europe, the population is even more concentrated in cities. More than half of Iceland's people live in the capital, Reykjavik. The capitals of Denmark, Norway, Sweden, and Finland have more people than any cities in their countries and are the cultural centers as well. In Northern Europe, most people live in southern parts of the countries because of the milder climates.

People and Cultures

Guiding Question *Why are most people in Northern Europe Protestant, whereas most people in Southern Europe are Catholic?*

The populations of the Northern European countries are fairly **homogeneous**, or similar. The original settlers of these lands were the Sami people, or Lapps. Many Sami live by fishing and hunting. Most live in northern Sweden, Norway, and Finland. Iceland was first settled by Celts and later conquered by Norway. The bulk of the population is a blend of the two ethnic groups.

The languages spoken in Denmark, Sweden, and Norway are from a common German language base. Today, Danish, Swedish, and Norwegian are very similar. Finnish is more like the Hungarian and Estonian languages.

Northern and Southern Europe

Lesson 3: Life in Northern and Southern Europe, *continued*

The populations of Southern European countries are also fairly homogeneous. Greece has small populations of Turks, Albanians, Macedonians, and Rom (gypsies). Most people speak Greek.

Most people in Italy speak Italian. However, minority groups in northern Italy speak German, French, or Slovenian. The Italian language has many **dialects**, or regional variations.

Most people in Spain speak a form of Spanish called Castilian. Other dialects are spoken as well, such as Catalan and Galician, which is related to Portuguese.

The Protestant Reformation was more successful in Northern than in Southern Europe. The table below shows the common religions in the regions.

Region	Religion
Northern Europe	More than 75% are Lutheran, with small percentages Catholic or Muslim
Spain, Portugal, Italy	Most are Roman Catholic; Islam growing due to immigrants
Greece	98% are Greek Orthodox

Italy has made many contributions to art in painting, sculpture, and architecture. The genius of Italian Renaissance and Baroque art can be seen in palaces, churches, and paintings in cities such as Rome and Florence. Italian operas are among the most popular in the world. Italy is one of the world centers for architecture and fashion today.

Spanish music has been influential all over Europe and the Americas. Spain promoted the guitar as a serious musical instrument. In Northern Europe also, composers made important contributions to classical and pop music.

One of Northern Europe's most notable achievements is its literacy rate. In Norway, Sweden, and Denmark, the rate is nearly 100 percent. Schools have strong support from the government and most schooling is free. People in Northern Europe pay high taxes, but they receive many public services and social welfare benefits.

Winter sports, such as skiing, are popular in Northern Europe. Soccer is popular in Southern Europe. Basketball is also common in Spain and Greece. And in Spain, bullfighting is still popular.

Northern and Southern Europe are affected by the spread of popular culture and music. American fast-food chains, television programs, and films are well liked throughout these regions.

Determining Word Meaning

5. What word describes the regional variations of a language?

_____

Marking the Text

6. Highlight Italy's contribution to art and music in one color. Highlight Spain's contributions in another color.

Determining Cause and Effect

7. Why do you think Northern Europe's literacy rate is so high?

_____

_____

_____

Reading Progress Check

8. How is life in Southern Europe similar to life in the United States? How is it different?

_____

_____

_____

_____

_____

Lesson 3: Life in Northern and Southern Europe, *continued*

Marking the Text

9. Underline the passage that explains how Italy's economy improved after World War II.

Reading Progress Check

10. How does welfare capitalism work, and what advantages does it offer the people of Northern Europe?

_____

_____

_____

_____

_____

Issues in Northern and Southern Europe

Guiding Question *How do the financial problems of one country in Southern Europe affect other European countries?*

In Northern European countries, the standard of living is high, but so are taxes. Northern Europeans practice a form of **welfare capitalism**. In this system, the government uses tax money to provide a variety of services, such as health care and education, to all citizens.

After World War II, Italy had one of the weakest economies in Europe. Since then, it has built a strong industrial base. This base is stronger in the north than the south of the country.

In 1992, a group of European countries signed a treaty creating the European Union (EU). The EU helps ease conflicts among member countries, but there are still problems. For instance, when many immigrants move to a country, fear of cultural differences can cause conflict and violence. A financial crisis swept the world in 2008. Europe had a **recession**, or a period of slow economic growth or decline. The three biggest banks in Iceland failed. Spain and Greece were hit hard. This affected the economies of other countries that were joined to them through the EU.

Another issue surrounds two countries that are not in the EU, Iceland and Norway. Whaling was an important industry for both countries. However, as the whale populations shrank, a ban was put in place. The ban has been lifted several times but currently the International Whaling Commission is watching these countries.

Writing

Check for Understanding

1. Informative/Explanatory How does the economic system of Northern Europe support its high literacy rate?

_____

_____

2. Informative/Explanatory In what ways are the people of Northern Europe similar?

_____

_____

_____

networks

Eastern Europe and Western Russia

Lesson 1: Physical Geography

ESSENTIAL QUESTION

How does geography influence the way people live?

Terms to Know

upland the high land away from the coast of a country
steppe a partly dry grassland often found on the edge of a desert
balkanization conflict among ethnic groups within a region
brackish water that is somewhat salty
reserves estimated amount of a resource in an area

Where in the World: Eastern Europe and Western Russia

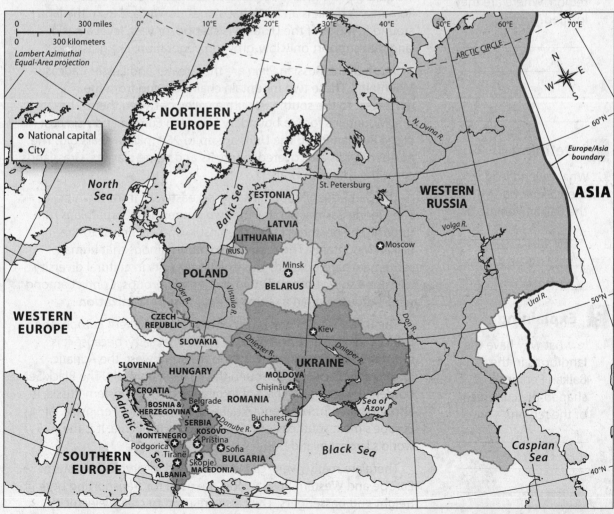

Lesson 1: Physical Geography, *continued*

Marking the Text

1. Read the text on the right. Underline the sentence that tells where the border between Europe and Asia is located.

? Locating

2. Two basins are part of the geography of the region. Where are they located?

_____

_____

_____

_____

Aᵇᒼ Defining

3. What helped lead to the *balkanization* of the Balkan Peninsula?

_____

_____

_____

? Explaining

4. In what way have the landforms in the Balkan Peninsula shaped the cultures of the region?

_____

_____

_____

Landforms and Waterways

Guiding Question *In what way have the landforms in the Balkan Peninsula shaped the cultures of that region?*

Eastern Europe includes 10 countries in the north and 11 on the Balkan Peninsula. Russia is a huge country, extending through 11 time zones in Europe and Asia. Western Russia is the portion that lies within Europe. The region rests mostly on a group of plains. The largest is the Russian Plain. It begins in Belarus and Ukraine and stretches east about 1,000 miles (1,609 km). It rises to form the central Russian **upland**. This is an area of high elevation. To the east are the Ural Mountains. Beyond that is the west Siberian plain.

The Northern European Plain includes Poland and extends into parts of Western Europe. South of it is the Hungarian Plain, which includes parts of many countries. The Transylvanian Basin is in Romania. A basin is an area that slopes downward from the land around it. Much of the Ukraine is **steppe**, or vast, level areas of land that support only low-growing vegetation, like grasses.

South of the Russian Plain are the Greater and Lesser Caucasus Mountains. These two mountain chains extend from the northwest to the southeast with a valley between them. The Ural Mountains form a boundary between Europe and Asia east of the Russian Plain. The Urals are up to 250 million years old. The northern Urals are covered in forests and some glaciers. Grasslands cover the southern Urals.

The Carpathian Mountains extend eastward from the Alps. The Vienna Basin separates the two ranges. The Carpathians run through the northern Balkan Peninsula and are linked to the Balkan Mountains. The region is so mountainous that human settlements are isolated. This isolation results in cultural diversity. It is also a source of conflict among ethnic groups. Conflict among ethnic groups within a region is known as **balkanization**.

The Baltic Sea lies northwest of Russia and Eastern Europe. The Baltic is shallow and **brackish**, or somewhat salty, because it is seawater mixed with river water. In the southwest, the Adriatic, Ionian, and Black Seas surround the Balkan Peninsula. The Black Sea borders the southern coast of Ukraine and southwestern Russia. It also separates Turkey from Ukraine and the Balkan Peninsula. At Europe's most southeastern point is the Caspian Sea. It is the world's largest inland body of water.

There are many rivers, canals, lakes, and reservoirs in Eastern Europe and Western Russia. Most are used for transporting both freight and passengers. They also are used to create electricity.

Eastern Europe and Western Russia

Lesson 1: Physical Geography, *continued*

Rivers of Eastern Europe and Russia		
River	**Location**	**Use**
Volga	Russia	freight and passenger traffic, hydroelectric power
Dnieper	Russia, Belarus, Ukraine	hydroelectric power, irrigation
Dniester	Carpathian Mountains to the Black Sea	freight and passenger ships
Danube	southwestern Germany to the Black Sea	transportation, hydroelectric power, fishing, irrigation

Climates

Guiding Question *How does climate affect plants that are grown and harvested in Eastern Europe and Western Russia?*

Several types of climate are found in Eastern Europe and Western Russia. Much of the region has a humid continental climate. These areas have mild or warm summers and long, cold winters. In places such as Croatia, Serbia, and Bulgaria, summers are hotter, and winters are similar to those farther north. Albania and Macedonia have a more Mediterranean climate. Summers are hot and dry, and winters are mild to cool and rainy.

Russia's far north has a subarctic climate. Winters are very cold, with temperatures as low as −40°F (−40°C). Summers are short and cool, although temperatures can reach 86°F (30°C). Farther north is Novaya Zemlya. This is an archipelago consisting of two large islands and several small islands. The climate is polar, and a large part of it is covered in ice year-round. A small number of herders and fishers inhabit the southern island.

Natural Resources

Guiding Question *What are three important challenges to the development of resources in Eastern Europe and Western Russia?*

Russia is by far the largest country in the world. However, only about one-sixth of its land is suitable for agriculture. Most agricultural land is in a fertile triangle extending from the Baltic Sea to the Black Sea. Farmers grow crops such as wheat, oats, and barley.

✎ **Marking the Text**

5. Review the table on the left. Highlight the names of rivers that provide hydroelectric power to the region.

☑ **Reading Progress Check**

6. How is the location of the Black Sea strategic to the region?

_____

_____

_____

❓ **Explaining**

7. Why do grasses and low, scrubby bushes form the primary vegetation on the archipelago of Novaya Zemlya?

_____

_____

☑ **Reading Progress Check**

8. What area in Eastern Europe or Western Russia has a climate most similar to where you live?

_____

_____

Lesson 1: Physical Geography, *continued*

Marking the Text

9. Read the text on the right. Highlight the names of products of Russia's forest industry.

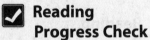

Reading Progress Check

10. Why are Russia's mineral industries so important to its economy?

_____

_____

_____

_____

More than one-fifth of the world's forests are in Russia. They cover an area almost the size of the continental United States. Lumber, paper, and cardboard are important products. Forests grow slowly because of the long, cold winters. Intense harvesting and slow growth rates threaten both forests and the forest industry. Wildfires in 2010 destroyed 37 million acres (15 million ha), further damaging Russia's forestry and agricultural industries.

The countries of this region also have many mineral resources. Most of Russia's vast coal, oil, and natural gas **reserves** are in Siberia. Reserves are the estimated total amount of a resource in an area. Russia's coal and rich deposits of iron ore fuel the country's steel industry. Machines made from steel are used to build Russia's automobiles, railroads, ships, and many consumer products.

Poland's mineral resources include aluminum, coal, copper, lead, zinc, and sulfur. Romania has rich coal deposits and oil in the Black Sea. Hydroelectric and thermal power plants provide energy. There are also copper and bauxite, the raw material for aluminum.

Russia's fishing industry is important to the country's economy. Many of Russia's lakes and rivers are used for freshwater fishing. Romania's fishing industry is concentrated in the southeastern part of the country. The Danube River supplies the most fish.

Writing

Check for Understanding

1. **Informative/Explanatory** What role does the Danube River play in Eastern Europe?

_____

_____

_____

2. **Informative/Explanatory** Which energy sources are important in Eastern Europe and Western Russia?

_____

_____

_____

_____

netw⊙rks

Lesson 2: History of the Regions

ESSENTIAL QUESTION
How do governments change?

Terms to Know

czar title given to an emperor of Russia's past
serf a farm laborer who could be bought and sold along with the land
genocide the mass murder of people from a particular ethnic group
communism a system of government in which the community or the state owns all property
collectivization a system in which all farmland is owned and controlled by the government

---

When did it happen?

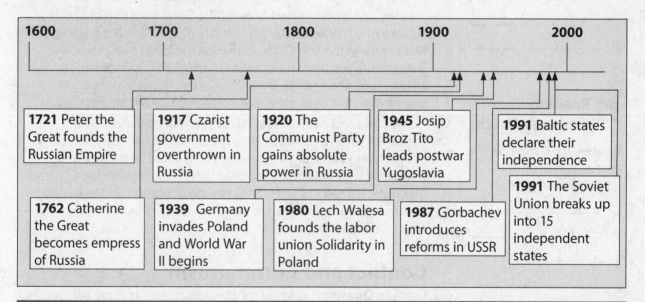

1600	1700	1800	1900	2000

1721 Peter the Great founds the Russian Empire

1917 Czarist government overthrown in Russia

1920 The Communist Party gains absolute power in Russia

1945 Josip Broz Tito leads postwar Yugoslavia

1991 Baltic states declare their independence

1762 Catherine the Great becomes empress of Russia

1939 Germany invades Poland and World War II begins

1980 Lech Walesa founds the labor union Solidarity in Poland

1987 Gorbachev introduces reforms in USSR

1991 The Soviet Union breaks up into 15 independent states

Early History

Guiding Question *How did Peter I and Catherine II change Russia?*

Many ethnic groups settled in Eastern Europe and Western Russia long before modern national borders were set. Most of the people are Slavs. Slavs are an ethnic group that includes Poles, Serbs, Ukrainians, and other Eastern Europeans. Originally from Asia, they settled in what is now Ukraine and Poland.

In the A.D. 800s, Slavic groups formed an empire that covered much of central Europe. Other Slavs settled in the Balkans, coming under the rule of the Ottoman Empire. Another Slav group settled in the forests and plains of what are Ukraine and Belarus today.

Marking the Text

1. Read the text on the left. Circle the locations where Slavic people settled in Eastern Europe.

Lesson 2: History of the Regions, *continued*

Marking the Text

2. Read the text on the right. Highlight the major ethnic groups that live in Western Russia.

Determining Word Meaning

3. Use the word *serf* in a sentence.

_____

_____

_____

Reading Progress Check

4. How did Catherine the Great expand the Russian empire?

_____

_____

_____

_____

Describing

5. What event led to the outbreak of World War I?

_____

_____

_____

_____

There they formed a union of city-states called Kievan Rus. Its leaders controlled the area's trade, using Russia's western rivers as a link between the Baltic Sea and the Black Sea.

The people of Western Russia are a mix of traditional ethnic groups. Ethnic Russians make up the majority of the population, but there are also Magyars, Romanians, Slavs, Ukrainians, and many others. Russian Slavs have dominated the country's politics and culture. Most Slavs practice Eastern Orthodox Christianity, which was brought to Russia from the eastern Mediterranean area.

During the later 1200s, Mongols from Central Asia invaded Russia. For the next 250 years, they controlled most of Russia. Near the end of the Mongol reign, the princes of Muscovy (now Moscow) came to power. Ivan IV defeated the Mongols and declared himself the ruler, or **czar**.

Later, a czar now known as Peter the Great tried to make Russia a major power. After his death in 1725, there was a string of weak czars. In the late 1700s, Catherine the Great came to power. She encouraged the development of education, journalism, architecture, and theater. During her reign, Russia expanded its empire and took over the entire northern coast of the Black Sea.

The czars and nobles enjoyed rich, comfortable lives. At the bottom of society, however, were great masses of people. Most of them were **serfs**, or farm laborers who could be bought and sold along with the land. They lived hard lives. In 1861 Czar Alexander II freed the serfs. They had no education and few ways to earn a living. Some went to cities, where they worked long hours for low wages.

Conflict and Communism

Guiding Question *How did the Russian Communist Party plan to transform Russia into an industrial giant?*

In the early 1900s, the threat of war had been brewing for many years. The major powers had already formed alliances. War broke out when a Bosnian terrorist named Gavrilo Princip assassinated Archduke Francis Ferdinand of Austria-Hungary on June 28, 1914, in Bosnia. By August, nearly all of Europe was involved in what would come to be known as World War I.

Allied Powers	Central Powers
Great Britain	Austria
France	Germany
Russia	Ottoman Empire

Eastern Europe and Western Russia

Lesson 2: History of the Regions, *continued*

At first the Russian people supported the war. However, military failures, high casualties, and food shortages turned public opinion against the war and the czar. Russia encouraged Armenians in Turkish-controlled lands to help them. The Turks responded by deporting 1.75 million Armenians to Syria and Mesopotamia. About 600,000 Armenians starved or were murdered by Turkish soldiers and police. The mass murder of vast numbers of an ethnic or cultural group is called **genocide**.

In 1917 food shortages in Russia triggered riots in the capital. Soldiers deserted and joined civilians in protests against the war. Czar Nicholas was forced to step down. A new government was formed, but it could not hold onto power. By the end of the year, a group of revolutionaries called the Bolsheviks had seized control of the government. The Bolsheviks had strong support all over Russia.

Inspired by the writings of Karl Marx, the Bolsheviks made Russia a communist state. **Communism** is an economic system that says all property should belong to the community or state, not to private individuals. The Bolsheviks became the Russian Communist Party and took control of all land and industry. Their leader, Vladimir Lenin, became the first premier of the new Union of Soviet Socialists Republics, or Soviet Union.

After Lenin's death in 1924, Joseph Stalin became leader of the Soviet Union. Stalin used terror and brute force to make the country a communist dictatorship. He forced the **collectivization** of all agriculture, so that all farmland was owned and controlled by the government. Peasants and landowners protested, especially in Ukraine. Their clash with the government led to a famine that killed millions. By the early 1930s, the Soviet Union was on its way to becoming an industrial giant. Stalin, however, wanted to spread communism throughout the world.

In 1941 Nazi Germany invaded the Soviet Union, drawing the country into World War II. The Soviets joined with Great Britain and the United States to defeat the Germans. When the war ended, the Soviet army already occupied Czechoslovakia, Poland, Romania, Hungary, and Bulgaria. At first Stalin agreed to allow these countries to hold elections. However, he went back on this promise and soon installed communist governments.

Germany was split in two. The U.S. and Great Britain set up a democracy in West Germany. East Germany became a communist state. Countries under Soviet rule came to be known as satellite countries. This meant that they were under the economic and political control of a more powerful country.

Marking the Text

6. Read the text on the left. Underline the sentence that tells why the Russian people turned against the czar.

Making Connections

7. What was the relationship between the Bolsheviks and the Communist Party?

_____

_____

_____

Analyzing

8. Why did landowners protest the *collectivization* of agriculture?

_____

_____

_____

Reading Progress Check

9. How did the USSR come to control most of Eastern Europe?

_____

_____

_____

networks

Lesson 2: History of the Regions, *continued*

Marking the Text

10. Read the text on the right. Underline the sentences that explain what NATO and the Warsaw Pact were.

? Identifying

11. What was Solidarity?

_____

_____

_____

⚙ Drawing Conclusions

12. How did the Soviet Union's policy of *glasnost* affect the governments of Eastern Europe?

_____

_____

_____

⚙ Sequencing

13. Create a timeline for the collapse of Soviet domination of Eastern Europe.

_____

_____

_____

_____

_____

The Regions in the Modern Era

Guiding Question *How is a "cold war" different from other kinds of war?*

The Cold War was the rivalry and conflict between two superpowers—the Soviet Union and the United States—and their allies. The two countries competed for world influence and power.

Both superpowers built destructive weapons. They also used other means, such as military and financial aid and the threat of force to carry out the struggle. At times, nuclear warfare seemed to be inevitable. Each side built alliances. The United States and its allies created the North Atlantic Treaty Organization (NATO) in 1949. NATO members agreed to respond as a group if any member of was attacked.

In 1955, the USSR responded by creating a new alliance with its satellites. Albania, Bulgaria, Czechoslovakia, East Germany, Hungary, Poland, and Romania joined the USSR in the Warsaw Pact.

The two superpowers came close to war during the Cuban Missile Crisis. In October, 1962, the Soviet Union began sending missiles to Cuba. The U.S. president sent the Navy to stop the Soviet ships. It seemed like there might be war. Eventually the Soviet leader agreed to remove the missiles. One crisis was avoided. However, another crisis was beginning in Eastern Europe.

In 1968, Alexander Dubček, the head of the Communist Party in Czechoslovakia, announced sweeping reforms. The Czech people welcomed the reforms. However, Soviet troops invaded Czechoslovakia and removed Dubček from power.

In 1980, dozens of Polish trade unions joined to form Solidarity. Solidarity used strikes to put pressure on the government. When the government declared Solidarity illegal, it became an underground organization.

Mikhail Gorbachev came to power in the Soviet Union in the 1980s. He announced new policies. *Glasnost*, or "openness," was an attempt to give the people in the USSR and its satellites more social and political freedoms. *Perestroika*, or "restructuring," was an attempt to reform the Soviet economy.

Change came quickly in Eastern Europe. Solidarity was legalized in Poland in 1989, and in the next elections, the Communists were voted out of power. By 1990, Hungary, Czechoslovakia, Bulgaria, and Romania had new governments. The Berlin Wall separating West and East Berlin was torn down. By the next year, East and West Germany were reunited.

Eastern Europe and Western Russia

Lesson 2: History of the Regions, *continued*

In 1991 a group of Soviet officials staged a coup and arrested Gorbachev. His allies resisted, and the people protested. The military turned against the coup leaders. Gorbachev was released. Communist control of the USSR was over. By the end of the year, the Soviet Union had dissolved. All its republics became independent countries.

Ethnic tensions flared in the Balkan Peninsula. In the early 1990s, disputes among ethnic groups tore Yugoslavia apart. Kosovo, which was part of Serbia, has a mostly Albanian Muslim population. Many people in Kosovo wanted to be free of Serbian control. When armed rebellion broke out in 1998, Serbs responded with military force. NATO intervened to end the bloodshed, and the United Nations began governing Kosovo. Kosovo declared independence in 2008, but Serbia and its allies refused to recognize this.

Countries Created from the Former Yugoslavia	
• Slovenia	• Macedonia
• Bosnia	• Herzegovina
• Serbia	• Montenegro
• Croatia	

Analzying

14. Why do you think the Soviet Union dissolved when Communist control ended?

_____

_____

Reading Progress Check

15. How did *glasnost* and *perestroika* affect the USSR?

_____

_____

_____

Writing

Check for Understanding

1. Argument Which of the two world wars do you think had a greater effect on the course of history in Western Russia and Eastern Europe? Why?

_____

_____

_____

2. Informative/Explanatory How was the Cold War different from other kinds of war?

_____

_____

_____

Eastern Europe and Western Russia

Lesson 3: Life in Eastern Europe and Western Russia

ESSENTIAL QUESTION

How does geography influence the way people live?

Terms to Know

inflation an increase in the price of goods and services

oligarch a member of a small group of people who control the government

devolution the process by which a large, centralized government gives power away to smaller, local governments

---

What Do You Know?

In the first column, answer the questions based on what you know before you study. After this lesson, complete the last column.

Now...		Later...
	How has Russia changed since the collapse of the Soviet Union?	
	Why did the Slavs develop different languages and cultures?	
	What has happened to the former Soviet satellites since the collapse of the USSR?	

? **Describing**

1. Describe the transfer of industry to private ownership in post-Soviet Russia.

   _____

   _____

   _____

   _____

People and Places

Guiding Question *What were some of the challenges Russia faced after the fall of the Soviet Union?*

Russia faced enormous challenges after the collapse of the USSR. Industry had centered on military hardware and heavy industrial machinery. The country was not prepared to produce the consumer goods that fuel free markets. **Inflation**, or the rise in prices for goods and services, increased, while production slowed. The transfer of industry to private ownership did not benefit most Russians. The new owners of industry were wealthy people with government connections called **oligarchs**. An oligarch is a member of a small group of people who control the government and use it to further their own goals.

Eastern Europe and Western Russia

Lesson 3: Life in Eastern Europe and Western Russia, *continued*

The Soviet government kept tight control over the USSR's many ethnic groups. Now some groups want to form their own countries. Among them are the Chechens, who live in Chechnya, in southern Russia. The region has many oil reserves that help fuel major Russian cities. Russian troops fought to keep Chechnya part of Russia. When they pulled out in 1996, the rebellion was still not over. President Boris Yeltsin was blamed for being unable to solve the problems. He resigned in late 1999.

Vladimir Putin was elected president in 2000. Putin had been an officer in the KGB, the country's secret police. He launched reforms to reduce the power of the oligarchs and encourage economic development. He dealt harshly with those who opposed him, and was reelected in 2004. After a term as prime minister, Putin was reelected as president in 2012. He soon passed laws restricting demonstrators, Internet access, and free speech.

The two largest cities in Western Russia are Moscow and St. Petersburg. Moscow is the political, cultural, educational, scientific and religious capital of Russia. St. Petersburg was founded by Peter the Great in 1703. The biggest population centers in Eastern Europe are the capital cities. They are also centers of national culture. Most Eastern European people live in urban areas. The urban population of Eastern European countries varies from 50 percent to 70 percent of a country's total population.

During the industrial age, people began moving from the region to Western Europe and North America. That trend continues. Eastern Europeans moved to escape political oppression and to seek better economic opportunities.

People and Cultures

Guiding Question *How did geographical barriers affect the development of Slavic culture in Eastern Europe?*

Originally, most Slavs spoke one language. As they settled different parts of Eastern Europe, geographical barriers separated them. They developed distinct languages and cultures. Most Slavs belong to one of three categories—East Slavs, West Slavs, and South Slavs.

East Slavs	West Slavs	South Slavs
Russia, Ukraine, Belarus	Poland, Slovakia, Czech Republic, parts of Germany	Bulgaria, Balkans

✏ Marking the Text

2. Highlight the names of two men who have served as presidents of Russia.

⚙ Determining Cause and Effect

3. What caused Boris Yeltsin to resign?

_____

_____

✔ Reading Progress Check

4. Why have so many Eastern Europeans emigrated to other parts of Europe or to the United States?

_____

_____

_____

_____

❓ Describing

5. How did geographical barriers affect the development of Slavic culture in Eastern Europe?

_____

_____

_____

_____

netw☉rks

Lesson 3: Life in Eastern Europe and Western Russia, *continued*

Marking the Text

6. Read the text on the right. Highlight the names of religions that are dominant in Eastern European countries.

Defining

7. What does the word *devolution* refer to?

_____

_____

_____

Reading Progress Check

8. How has daily life changed in Russia since the fall of the USSR?

_____

_____

_____

_____

_____

Marking the Text

9. Highlight the names of Eastern European countries where a large percentage of the population works in agriculture.

Russia has more than 120 ethnic groups, but almost 80 percent of its people are ethnic Russians. For example, the people of Albania are a distinct ethnic group that has been in the region for about 4,000 years. Their language is the last survivor of an entire Indo-European language group.

In some countries in Eastern Europe, a large percentage of the population does not practice any religion. In most of the region, however, Soviet repression strengthened religious faith. The dominant religion in most countries is the Eastern Orthodox Church. In several countries, the majority of people are Roman Catholics. The majority of Albanians are Muslim. Most countries have minority populations of Muslims, Roman Catholics, Eastern Orthodox, Protestants, and Jews.

In the 1800s and early 1900s, Russians produced some of the most important cultural works in Europe. People in the countries of Eastern Europe are proud of the art produced by their people. In many cases, those works are an important symbol of their national character. Eastern European composers often celebrate their traditional music by including it in their compositions.

Western popular culture has had a huge impact on the region. American films and television, as well as rock and pop music from the US and Europe are well known. Young artists are creating international popular culture with an Eastern European twist.

The collapse of the USSR brought about **devolution** in Russia and Eastern Europe. Devolution occurs when a strong central government surrenders its powers to local authorities. One result of this change is a return to national traditions and identity. Another is the rising influence of international popular culture. This can cause a generation gap between people who grew up in the communist era and those who have lived all or most of their lives in the post-Soviet era.

Issues in Eastern Europe and Western Russia

Guiding Question *What are the economic advantages and disadvantages of Eastern Europe's location between continents?*

Western Russia and Eastern Europe cover a vast area. The region has many different landforms, soil, natural resources, climate, and economies. These factors combine to determine how people earn a living. Albania, Romania, Serbia, and Bosnia and Herzegovina have many people working in agriculture.

networks

Lesson 3: Life in Eastern Europe and Western Russia, *continued*

Russia is a leading oil and natural gas producer. It is also a major supplier of iron ore and other metals. About 1 million people work in Russia's forestry industry. The majority of people in Russia and Eastern Europe now work in the service industries, providing services to individuals and businesses.

Many of the industries in Eastern Europe fell on hard times after the collapse of the Soviet Union. Big losses led to high unemployment, especially in the Balkans. Many Eastern European countries have joined the European Union. The EU has tried to improve conditions for workers and create a strong European trading bloc that can compete effectively with the United States.

Russian culture has always been a mix of European and Asian influences. Today, oil and natural gas from oil fields in Siberia are delivered by pipeline to all of Russia, Eastern Europe, and beyond. Geographically and culturally, Russia plays a key role in the relationship between Europe and Asia.

There are still boundary disputes between Russia and the former Soviet republics. Even though fighting died down in Chechnya, there are still occasional outbreaks of violence.

The 2008 financial crisis hit Eastern Europe hard as the nations were trying to transform into free market economies. Joining the EU should have been a great benefit to their economies, but all members of the EU have suffered as a result of the financial crisis.

✎ Marking the Text

10. Read the text on the left. Highlight one way that Russia has kept some control over Eastern European countries.

☑ Reading Progress Check

11. What have been two setbacks in the economies of Eastern Europe since the collapse of the USSR?

_____

_____

_____

Writing

Check for Understanding

1. Informative/Explanatory What were some of the challenges Russia faced following the collapse of the Soviet Union?

_____

_____

_____

2. Argument What influence has Western popular culture had in Eastern Europe?

_____

_____

_____

East Asia

Lesson 1: Physical Geography of East Asia

ESSENTIAL QUESTION
How does geography influence the way people live?

Terms to Know
de facto actually; in reality
archipelago a group of islands
tsunami a giant ocean wave caused by an earthquake under the ocean floor
loess a fine-grained, fertile soil deposited by the wind

Where in the World: East Asia

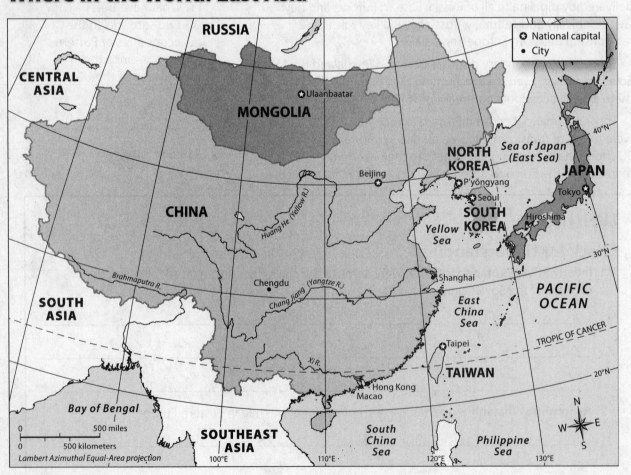

networks

East Asia

Lesson 1: Physical Geography of East Asia, *continued*

Landforms and Waterways

Guiding Question *What are the main physical features and physical processes in East Asia?*

East Asia is made up of six countries: China, Japan, Mongolia, North Korea, South Korea, and the de facto country of Taiwan. A **de facto** country is one that is not legally recognized. The region's largest country, China, is the world's fourth-largest country in land area.

Mainland East Asia, which includes China and Mongolia, can be divided into three subregions like steps. The highest step is the Plateau of Tibet. Much of it is more than 2.5 miles (4 km) above sea level. High mountains circle the plateau of Tibet. The Kunlun Shan range is on the north. On the south are the Himalaya, the tallest mountains in the world.

The middle step is north and east of Tibet and has lower mountains and plateaus. Much of the land to the north is desert or near desert. Land along the southern part is forested. There are deep canyons where the land descends from Tibet.

Low hills and plains form the third and lowest step, which covers most of the eastern third of China. Most Chinese people live on these plains.

In addition to the mainland, East Asia includes a large peninsula between the Yellow Sea and the Sea of Japan (East Sea). It is home to two countries, North Korea and South Korea. The peninsula is mountainous in the northeast. In the south and west, broad plains stretch between the mountains and the coast.

Japan is an **archipelago**, or chain of islands, along the eastern edge of the Sea of Japan. It is roughly 1,500 miles (2,414 km) long, and consists of four large islands and thousands of smaller ones. The islands of Japan are part of the Ring of Fire, which nearly encircles the Pacific Ocean. The islands were formed by volcanic eruptions millions of years ago. Mount Fuji, a beautiful, cone-shaped volcano, is a well-known symbol of Japan.

Japan is one of the most earthquake-prone countries in the world. When an earthquake occurs below or close to the ocean, it can cause a **tsunami**. This is a huge wave that gets higher as it approaches the coast. Tsunamis can wipe out coastal cities and towns.

Hundreds of miles southwest of Japan's main islands lies another large island, Taiwan. Like Japan, it was formed by volcanic activity. Mountains stretch the length of the island. On the western side of the island there is a gentler slope than on the steep eastern side. Broad plains spread across the western part of the island.

Defining

1. What makes Taiwan a *de facto* country?

   _____
   _____

? Identifying

2. What are the three geographic subregions of mainland East Asia?

   _____

   _____

   _____

   _____

? Explaining

3. How did Japan's *archipelago* form?

   _____

   _____

Marking the Text

4. Read the text on the left. Highlight the names of the East Asian countries that are not found on the mainland.

Marking the Text

5. Underline the definition of a *tsunami*.

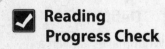
East Asia

Lesson 1: Physical Geography of East Asia, *continued*

Reading Progress Check

6. What are some ways the people of East Asia depend on rivers?

_____

_____

_____

Marking the Text

7. Read the text on the right. Highlight the major factors that affect climates in East Asia.

Contrasting

8. How do the climates of East Asia's island and peninsula areas differ from climates of the mainland areas?

_____

_____

_____

Reading Progress Check

9. How do the Himalaya affect the climate of the Plateau of Tibet?

_____

_____

_____

Four seas sit along the eastern edge of East Asia. The South China Sea lies between southeastern China and Taiwan. The East China Sea lies between China and Japan. In the north, it meets the Yellow Sea, which is shaped by the Korean Peninsula and the northeastern coast of China. Farther north, Japan, the Korean Peninsula, and the Asian mainland nearly surround the Sea of Japan (East Sea).

East Asia's two most important rivers are the Huang He (Yellow River) and the Chang Jiang (Yangtze River). The Huang He gets its name from yellow-brown silt called **loess**. Silt deposited by floods has created a broad, fertile plain that has some of China's best farmland. These floods have also caused much damage and loss of life. The Chang Jiang is the longest river in Asia and China's principal waterway. It also provides water for rice farming.

Japan's major rivers are short, steep, and swift. Most of them generate hydroelectric power. The main rivers of the Korean Peninsula flow from inland mountains toward the Yellow Sea. North Korea's longest river, the Yalu, forms the country's border with China.

Climate

Guiding Question *What are the main factors that affect climate in different parts of East Asia?*

Climates in East Asia vary greatly because of several factors:

- Latitude: The region spans a great distance from north to south.
- Elevation: Two areas at the same latitude can have very different climates if one is higher than the other.
- Air masses: Cold, dry, polar air spreads from northern Asia in colder months. Warm, moist, tropical air spreads northward and eastward from the Pacific Ocean in warmer months.

Southeastern China is hot and rainy much of the year, with lush vegetation. To the north, there is more seasonal variation. Taiwan, Japan, and the Korean Peninsula are generally wetter and have milder temperatures than mainland areas at the same latitudes.

The climate in Mongolia and north-central and northwestern China is dry. Winters are bitterly cold. This region includes the Gobi Desert and the Taklimakan desert, as well as treeless grasslands.

The Plateau of Tibet in southwestern China also has a dry climate. The Himalaya block moist air flowing northward from the Indian Ocean. Because of the elevation, the plateau is cold and windy throughout the year.

networks

East Asia

Lesson 1: Physical Geography of East Asia, *continued*

Natural Resources

Guiding Question *What mineral resources are most abundant in East Asia?*

China holds the greatest share of the region's resources. Japan is one of the world's leading industrial countries, but has few mineral resources. It must import many raw materials. Taiwan, another major industrial country, also has limited resources and must import minerals to meet demand.

Mineral Resources in East Asia	
China	tin, lead, zinc, iron ore, tungsten, other minerals
Japan	coal, copper, some iron ore, other minerals

The largest deposits of fossil fuels are in China. China is the world's largest producer of coal. It also has large oil and natural gas reserves under the South China Sea and in the Taklimakan desert in the far west. Despite these resources, China still cannot meet all of the energy needs of its growing economy. Both China and Japan use hydroelectric dams to produce electricity.

Eastern China was once covered by forests, but people cut trees down for heating, building, and to create farmland. Today, forests cover less than one-sixth of the country. More than half of Taiwan is covered in forests. However, much of the forested land is protected, so Taiwan must import wood. Almost two-thirds of Japan is forested. Logging is limited because the Japanese consider many forest areas to be sacred. In the Korean Peninsula, many trees have been cleared for farmland. About three-fourths of North Korea is forested.

Marking the Text

10. Read the text on the left. Highlight the names of two major industrial countries in East Asia that must import most of their raw materials.

Reading Progress Check

11. Why is it necessary for people in Taiwan and Japan to import wood products?

_____

_____

_____

_____

_____

_____

Writing

Check for Understanding

1. **Informative/Explanatory** Summarize the way mineral resources are distributed among the countries of East Asia.

_____

_____

_____

East Asia

Lesson 2: History of East Asia

ESSENTIAL QUESTIONS

What makes a culture unique? • How do cultures spread?

Terms to Know

dynasty a line of rulers from a single family that holds power for a long time

shogun a military leader who ruled Japan in early times

samurai a powerful, land-owning warrior in Japan

sphere of influence an area of a country where a single foreign power has been granted exclusive trading rights

communism a system of government in which the government controls the ways of producing goods

When did it happen?

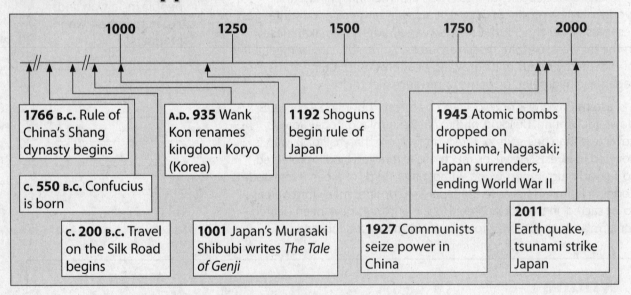

| 1000 | 1250 | 1500 | 1750 | 2000 |

1766 B.C. Rule of China's Shang dynasty begins

c. 550 B.C. Confucius is born

c. 200 B.C. Travel on the Silk Road begins

A.D. 935 Wank Kon renames kingdom Koryo (Korea)

1001 Japan's Murasaki Shibubi writes *The Tale of Genji*

1192 Shoguns begin rule of Japan

1927 Communists seize power in China

1945 Atomic bombs dropped on Hiroshima, Nagasaki; Japan surrenders, ending World War II

2011 Earthquake, tsunami strike Japan

Abc Defining

1. How did a new *dynasty* form?

_____

_____

_____

Early East Asia

Guiding Question *What important inventions from East Asia spread across the rest of the world?*

Chinese civilization is over 4,000 years old. For many centuries until the early 1900s, emperors or empresses ruled China. A **dynasty**, or line of rulers, held power until it was overthrown. Then a new leader would start a new dynasty. Under the dynasties, China developed an advanced culture and conquered neighboring lands.

networks

East Asia

Lesson 2: History of East Asia, *continued*

The Chinese tried to keep out foreign invaders. Most of its borders were natural barriers such as seas, mountains, and deserts. To the north, the Chinese built the Great Wall for protection, beginning about 2,200 years ago. Over the centuries, it grew to stretch thousands of miles from the Yellow Sea in the east to the deserts in the west. It remains in place today.

A great thinker named Confucius believed in the importance of the family. He thought a ruler should lead as though he were the head of a family. He founded the belief system called Confucianism. Another thinker, Laozi, founded a belief system called Daoism. It taught that people should live in harmony with nature. A third belief system, Buddhism, came to China from India.

Under the Han dynasty (202 B.C.–A.D. 220), the arts and sciences flourished. The process of making paper was invented, and officials began keeping paper records. Han rulers encouraged trade along the Silk Road. This was a caravan route that stretched between China and Southwest Asia, into Europe and South Asia. The Chinese sent silk, tea, spices, paper, and fine porcelain west as far as the Mediterranean in exchange for wool, gold, and silver.

New inventions changed life for the Chinese people and became important around the world.

Chinese Inventions	
Printing	made producing books faster and easier
Gunpowder	used in explosives and fireworks
Magnetic compass	helped sailors find direction at sea

Korea was settled by people from northern Asia. In the 1200s, Mongols invaded China and Korea. They were driven out at the end of the 1300s and a new Korean dynasty came to power. It stayed in power until modern times. In the A.D. 300s Buddhism spread to Korea from China. Later, Confucianism became the major religion.

Korean writing used Chinese characters, and Korean artists and writers were inspired by Chinese art and literature. Korean rulers adopted Confucianism as a basis of government. In some periods China provided military protection, but in others Koreans feared Chinese invasion.

The Japanese islands were settled by people from Korea and China. Close ties with China led to a flow of ideas and culture that transformed Japan. The Japanese used the Chinese calendar and the Chinese system of writing. They adopted Chinese technology. Buddhism spread to Japan from Korea. There it mixed with a Japanese religion called Shinto.

✎ Marking the Text

2. Read the text on the left. Highlight the names of three major belief systems that were important in China.

⚙ Drawing Conclusions

3. Besides physical goods such as silk and gold, what else might have been exchanged over the Silk Road?

_____

_____

_____

❓ Explaining

4. How did Buddhism spread across East Asia?

_____

_____

☑ Reading Progress Check

5. What are some ways in which China influenced Japan?

_____

_____

_____

_____

Lesson 2: History of East Asia, *continued*

Analyzing

6. What is the relationship between *samurai* and *shoguns*?

_____

_____

_____

_____

Marking the Text

7. Read the text on the right. Underline the sentence that explains what *spheres of influence* are.

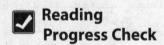

Reading Progress Check

8. How was Korea affected by Japanese expansion?

_____

_____

_____

_____

Marking the Text

9. Read the text on the right. Underline the sentences that describe results of China's Great Leap Forward.

Japan was ruled by emperors. Over time, they began to lose power, and landowning families set up a feudal system. Under this system, nobles gave land to lesser nobles in exchange for their loyalty and military service. At the bottom of the social ladder were the peasants who farmed nobles' estates in exchange for protection. By the 1100s, a military leader called the **shogun** held the real power in Japan. He was supported by landowning warriors called **samurai**.

Change in East Asia

Guiding Question *How did increased contact with the West influence the region?*

Until the 1500s, East Asia was isolated from the rest of the world. By the 1890s, European governments and Japan had claimed large areas of China as **spheres of influence**. These are areas of a country where one foreign power has exclusive trading rights. Anger against foreign invaders helped lead to a revolution in 1911. By 1927, a new government was formed by the Nationalist Party, led by Chiang Kai-shek. The new government was challenged by Mao Zedong, who believed in **communism**, a system in which the government controls all means of production. After years of civil war, the Communists won power in 1949.

Around 1542, a Portuguese ship heading for China was blown off its course and landed in Japan. It was followed by more traders and Christian missionaries. By the early 1600s, Japan's rulers decided to isolate Japan. In 1854 the United States pressured the Japanese to open their country to foreign trade. Japan set out to transform itself into a modern industrial and military power and began to create an empire. By 1940, Japan controlled Taiwan, Korea, parts of mainland Asia, and some Pacific islands. This expansion was one reason that Japan fought the United States and its allies in World War II.

Modern East Asia

Guiding Question *What conflicts divided East Asian countries?*

After 1949, China became "two Chinas." On the mainland, the Communist government took over businesses and industry. It also took land and created state-owned farms. In the late 1950s, China introduced the Great Leap Forward to increase industrial output. Cities grew fast as peasants left the land to work in factories. A drop in food production and natural disasters led to widespread famine.

networks

East Asia

Lesson 2: History of East Asia, *continued*

During China's Cultural Revolution in the 1960s, intellectuals and students were sent to the countryside to work in manual jobs. Mao hoped to get rid of cultural elements that did not support his idea of communism. After Mao's death in 1976, economic reforms have helped China become a rising global power.

Taiwan's government limited the freedom of the people. By 1970, however, its leaders had introduced democratic reforms and developed an economy based on capitalism. The island became an economic powerhouse.

"Two Chinas"	
People's Republic of China	*Republic of China*
• Communist government	• Nationalist government
• Located on mainland	• Located on Taiwan

Following World War II, Korea was divided. South Korea was supported by the United States. Communist North Korea had ties to China and the Soviet Union. War between the two broke out in 1950 when North Korea invaded South Korea. The war ended in 1953. South Korea has followed capitalism and built a strong economy. North Korea's economy is controlled by the government. Its people face many hardships because most resources go to the military.

After being defeated in World War II, Japan lost its overseas territories and military might. The government worked closely with businesses to plan the country's economic growth. The Japanese became leading producers of ships, cars, cameras, and computers. By the 1990s, Japan was a global economic power.

⚙ Contrasting

10. How do North Korea and South Korea differ?

_____

_____

_____

_____

☑ Reading Progress Check

11. Why did China's economy grow after the 1970s?

_____

_____

_____

_____

Writing
Check for Understanding

1. Informative/Explanatory Why did Europeans want access to China and Japan?

_____

_____

2. Informative/Explanatory What led to the creation of "two Chinas"?

_____

_____

_____

East Asia

Lesson 3: Life in East Asia

ESSENTIAL QUESTIONS
Why do people trade? • How does technology change the way people live?

Terms to Know
urbanization growth of a city into nearby areas
megalopolis a huge city or cluster of cities with an extremely large population
trade deficit occurs when the value of a country's imports is higher than the value of its exports
trade surplus occurs when the value of a country's exports is higher than the value of its imports

What Do You Know?
In the first column, answer the questions based on what you know before you study. After this lesson, complete the last column.

Now...		Later...
	How have East Asia's cities changed in recent decades?	
	What belief systems do East Asians practice today?	
	What have been the effects of rapid economic growth in East Asia?	

? Identifying

1. What actions has China taken to slow the rate of population growth?

   _____

   _____

   _____

   _____

The People
Guiding Question *What types of geographic areas in East Asia have the highest population densities?*

Most people in East Asia live crowded together in river valleys, basins, deltas, or on coastal plains. The lands and climates there are favorable to agriculture and industry. These are some of the most densely populated areas on Earth.

   China has had a large population for much of its history. By the middle of the 1900s, explosive population growth was causing many problems. In 1979, government policies encouraging families to have no more than one child helped slow China's growth. The 2010 census showed a population of 1.37 billion people. The east is much more densely populated than the west and northwest.

Lesson 3: Life in East Asia, *continued*

Population growth in other parts of East Asia also slowed at the end of the 1900s. Japan's low birthrate means that the average age of the population has increased. Nearly one-fourth of the population is 65 or older. Since the mid-1990s, Japan has encouraged more births. Soon the country could face a shortage of workers and have to allow more foreign workers into the country.

Throughout China's history, most of its people lived off the land as farmers. Economic reforms in the late 1970s, however, caused a surge of **urbanization**. Millions of peasants left their farms and moved to cities. Today, nearly half of the country's people live in cities. Shanghai is China's largest city. There are dozens of Chinese cities with population greater than 1 million.

Shanghai	11 million
Beijing	7 million
Hong Kong	5 million

Urbanization began earlier in other East Asian countries. In Japan, two-thirds of the people live in cities. The cities of Tokyo, Osaka, Nagoya, and Yokohama form a **megalopolis**, or supersized urban area, along the coast. Greater Tokyo, Japan's largest city, is home to 32 million people. As South Korea industrialized, more people moved to cities. Now, 83 percent of South Koreans live in urban areas. The capital, Seoul, has more than 10 million people. Across East Asia, the standard of living in cities is generally higher than that in rural areas.

Culture in East Asia

Guiding Question *What are some of the cultural differences among East Asian countries?*

In each East Asian country, most people are ethnically similar and speak the national language. The official language in China is Mandarin, but many dialects are spoken.

People practice many religions and belief systems. Buddhism is practiced throughout the region, often alongside other religions. The governments of China and North Korea limit religious practice, believing that religion has no place in a communist country. In recent decades this policy has been relaxed somewhat in China.

A number of art forms have long been popular in East Asia. Artists in China, Korea, and Japan paint the rugged landscapes of their countries. Their works reflect a special reverence for nature.

? Describing

2. How is Japan trying to compensate for the rise in the average age of its people?

_____

_____

_____

_____

Ꭺᵇᴄ Defining

3. What caused China's rapid *urbanization*?

_____

_____

_____

_____

✎ Marking the Text

4. Read the text on the left. Highlight the names of the cities that make up Japan's *megalopolis*.

✔ Reading Progress Check

5. How can a country's growth rate influence its economy?

_____

_____

_____

_____

networks

Lesson 3: Life in East Asia, *continued*

? Identifying

6. What is calligraphy?

_____

_____

⚙ Analyzing

7. How have East Asian and Western popular culture influenced each other?

_____

_____

_____

✏ Marking the Text

8. Read the text on the right. Highlight the names of two American sports that are popular in East Asia in yellow. Highlight the names of two popular Asian sports in blue.

☑ Reading Progress Check

9. Why is religious activity limited in China?

_____

_____

_____

_____

Ceramics and pottery have been important parts of East Asian art since prehistoric times. Weaving, carving, and lacquerwork are also important. In China and Japan, calligraphy—the art of turning the written word into beautiful images—is considered one of the highest art forms. East Asians have strong literary and theatrical traditions. Both Japan and China are famous for their traditional forms of theater.

East Asians have also developed new forms of expression. Today, Japan is known for anime, a type of animation. Comic books and cartoons using this style are popular all over the world. South Korean "K-pop" music is popular with the young in Japan and other countries. It has its roots in dance and electronic music from the West.

People in East Asia enjoy many pastimes. Millions practice traditional martial arts such as tai chi and tae kwon do. American baseball is popular in Japan, Taiwan, and South Korea. American basketball is a top sport in China.

The family has traditionally been the center of social life in East Asia. In rural areas, different generations of one family may share a home. As more people have moved into urban areas, some traditional attitudes have begun to change. East Asian cultures place a high value on education. Teachers are respected, and children are expected to work hard.

Rice and noodles are staples in the diets of most East Asians, but cuisine varies widely in the region. There are many varieties of Chinese cooking. Japanese foods often include seafood and tofu. Mongolian meals often feature meat and dairy products.

Holidays are important in East Asia. Families gather to remember ancestors and celebrate on New Year's in Japan and China (where it is called Spring Festival). In Korea the Harvest Moon Festival is similar to Thanksgiving in the United States.

Current Issues in East Asia

Guiding Question *How do East Asian economies affect economies around the world?*

Recent rapid economic growth has transformed East Asia. Today, only the United States has a larger economy than China and Japan. With growth have come problems. In China, factories, coal-burning power plants, and the growing number of cars and trucks have led to dramatic increases in air pollution. Rapid urban growth has eaten up valuable farmland, and many cities face water shortages.

networks

East Asia

Lesson 3: Life in East Asia, *continued*

Japan has similar issues. Polluted air has caused acid rain and other problems, but Japan has tried to protect the environment. Earthquakes are a constant threat. In 2011 an earthquake in Japan killed thousands of people and damaged several nuclear power plants. It disrupted trade and manufacturing around the world.

Many of the goods manufactured in East Asia are shipped to the United States and Europe. Trade between China and the United States is not balanced. In 2010 the U.S. trade deficit with China rose to $273 billion.

trade surplus

occurs when a country *exports* more than it *imports*

trade deficit

occurs when a country *imports* more than it *exports*

Japan and China are both dealing with the challenges of population growth. China has fewer young adult workers as a result of its "one-child" policies, and the percentage of elderly people has grown rapidly.

Political differences are another challenge. Japan is in dispute with Russia over ownership of the Kuril Islands north of Japan. North Korea's efforts to build nuclear weapons have drawn harsh criticism from several countries. Both China and North Korea face questions about human rights. China continues to be pressured for its views on Tibet and Taiwan. China also faces a growing income gap between people in urban and rural areas.

? Identifying

10. Give two reasons why North Korea has faced international criticism.

_____

_____

_____

_____

_____

_____

✓ Reading Progress Check

11. How might an earthquake in Japan affect the economies of other parts of the world?

_____

_____

_____

Writing

Check for Understanding

1. **Informative/Explanatory** What effects might urbanization have on traditional family life in East Asia?

_____

_____

2. **Informative/Explanatory** What negative effects has economic growth had in East Asia?

_____

_____

networks

Lesson 1: Physical Geography of Southeast Asia

ESSENTIAL QUESTION
How does geography influence the way people live?

Terms to Know
insular an area consisting of islands
flora plant life
fauna animal life
endemic found only in one place or region

Where in the World: Southeast Asia

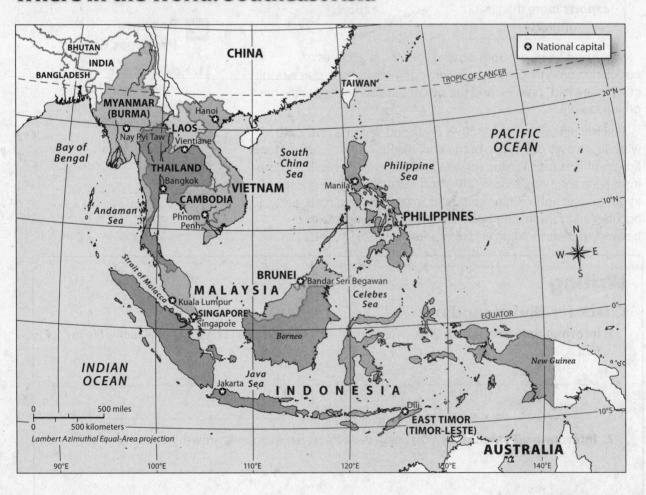

Southeast Asia

Lesson 1: Physical Geography of Southeast Asia, *continued*

Landforms and Resources

Guiding Question *How are the landforms of Southeast Asia's mainland different from the landforms of its islands?*

Southeast Asia can be divided into two parts. One is the mainland area. The other is an **insular** area, an area consisting of islands. There are 11 countries in Southeast Asia. Six of them are located at least partly on the mainland. Five countries are entirely on islands.

The mainland area sits at the southeastern corner of the Asian continent. It is mostly formed by two peninsulas. The larger is known as the Indochinese peninsula, or simply Indochina. The Malay Peninsula extends south from the Indochinese peninsula. The mainland is rugged and mountainous. The mainland countries are Myanmar (also known as Burma), Laos, Thailand, Vietnam, Cambodia, and Malaysia. Part of Malaysia is located on the island of Borneo.

The insular area is located where the Indian Ocean meets the Pacific Ocean. Thousands of islands stretch across miles of tropical waters. Many of the mountains on Southeast Asia's islands are volcanoes. The islands lie along the Ring of Fire.

The Malay Archipelago contains more than 24,000 islands. Singapore lies off the tip of the Malay Peninsula. Indonesia, the largest country in Southeast Asia, is located on more than 17,000 islands. Indonesia, East Timor, Brunei, and the Philippines are all part of the Malay Archipelago.

Four tectonic plates meet in Southeast Asia. As a result, many earthquakes and volcanic eruptions occur in this area. Volcanic eruptions can trigger deadly tsunamis. In 2004, a strong undersea earthquake produced huge waves that slammed into coastal areas of Southeast and South Asia. More than 230,000 deaths occurred.

Southeast Asia has many mineral resources. Tin, copper, lead, zinc, gold, and gemstones are all mined there. Indonesia and Malaysia have rich oil and natural gas reserves.

Bodies of Water

Guiding Question *Why does Southeast Asia have so many different seas?*

Bodies of water are important to Southeast Asia's geography and identity. The region covers about 5 million square miles (13 million sq km), but only a third of the area is land. Some of the world's busiest shipping lanes pass through Southeast Asia's seas.

 Marking the Text

1. Highlight the definition of *insular* in the text.

Activating Prior Knowledge

2. Define the word *peninsula*.

_____

_____

 Marking the Text

3. Underline the six mainland countries of Southeast Asia. Double underline the five countries located entirely on islands.

Defining

4. What is a tsunami?

_____

_____

Reading Progress Check

5. What effect has the region's location along the Ring of Fire had on the formation of the region?

_____

_____

Lesson 1: Physical Geography of Southeast Asia, *continued*

Marking the Text

6. In the text on the right, circle the names of oceans in the region. Draw a box around the names of the seas.

Activating Prior Knowledge

7. Why are river valleys productive farming areas?

_____

_____

Reading Progress Check

8. Why do you think Southeast Asia's longest rivers are found on the mainland and not on islands?

_____

_____

_____

Determining Cause and Effect

9. List three reasons why the vegetation of Southeast Asia is so rich.

_____

_____

_____

The Malay Peninsula and Indonesia's Sunda Isles are the boundary between two oceans. The Indian Ocean is in the southwestern area of the region. The Pacific Ocean is in the northeastern area. The two largest seas in the region are the South China Sea and the Philippine Sea. Both seas are part of the Pacific Ocean. The Andaman Sea is part of the Indian Ocean. Some of the busiest shipping lanes in the world pass through Southeast Asia's seas and their waterways.

Southeast Asia's longest and most important rivers are on the mainland. Most rain flows into one of five major rivers: the Irrawaddy, Salween, Chao Phraya, Mekong, and Red. Each river generally flows from the mountainous highlands in the north to the lowlands in the south. Then they empty into the sea.

The Irrawaddy River flows almost straight south through Myanmar's center. It is central to transportation. The river's fertile delta is important for farming. It drains into the Andaman Sea.

The Mekong River is Southeast Asia's longest river. It flows for about 2,700 miles (4,345 km) through or near Myanmar, Thailand, Laos, Cambodia, and Vietnam. Its drainage basin is twice the size of California. The river's enormous delta is one of the world's most productive agricultural regions.

Climate, Vegetation, and Wildlife

Guiding Question *In what ways does Southeast Asia's location shape its climate?*

Climates in Southeast Asia are generally hot and humid. Much of the region receives more than 60 inches (152 cm) of rain each year. These weather conditions and the variety of habitats in the region mean that there is a large quantity of plant and animal life.

Latitude and air currents are important factors in the region's climates. Most of the region lies within the Tropics. This is the zone that receives the hottest, most direct rays of the sun. Between November to March, monsoon winds blow across the region from the northeast to the southwest. These winds bring cooler weather to most of the mainland but heavy rains to the islands. From May to September, the monsoon winds switch directions. During these months, the mainland gets heavy rains. The islands have cooler, drier weather.

The waters surrounding Southeast Asia help moderate air temperatures. That means that the temperatures do not vary much. Elevation also affects weather conditions. Highland areas are generally cooler than the lowland areas.

networks

Lesson 1: Physical Geography of Southeast Asia, *continued*

Southeast Asia has four climate zones, which are shown in the table below.

tropical rain forest	• southern Malay Peninsula • southern Philippines • most of Indonesia
tropical monsoon climate	• northern Philippines • northern Malay Peninsula • coastal areas of the mainland
tropical savanna climate	• most inland areas
humid subtropical climate	• northernmost mainland

Weather in Southeast Asia can turn deadly. Tropical storms called typhoons form over the Pacific Ocean and sometimes make landfall. Winds can top 150 miles (241 km) per hour. The winds and pouring rains can destroy homes and buildings. Flooding can wipe out crops and kill large numbers of people.

Southeast Asia has a wide diversity of **flora**, or plant life. Much of the region is covered by tropical rain forests or forests of evergreen and deciduous trees. In coastal areas, forests of mangrove trees form a border between land and sea. The region's **fauna**, or animal life, is also very diverse. Many species of mammals, birds, fish, and insects are **endemic**. This means they are found nowhere else in the world. Unfortunately human activities such as logging, mining, and farming have reduced the habitat of many animals.

✎ Marking the Text

10. Highlight two reasons why typhoons are so dangerous.

🅰🅱🅲 Defining

11. Use the word *endemic* in a sentence.

_____

_____

_____

☑ Reading Progress Check

12. What are three factors that affect climate in Southeast Asia?

_____

_____

_____

Writing

Check for Understanding

1. Informative/Explanatory What types of landforms are found on mainland Southeast Asia? What types of landforms are found in the Malay Archipelago?

_____

_____

2. Narrative In which climate zone of Southeast Asia would you most like to live? Why?

_____

_____

Southeast Asia

Lesson 2: History of Southeast Asia

ESSENTIAL QUESTION
How does geography influence the way people live?

Terms to Know
sultan a king
plantation a large farm on which a single crop is grown for export
absolute monarchy a government in which one ruler has total control
constitutional monarchy a government in which a ruler must follow a constitution and laws

When did it happen?

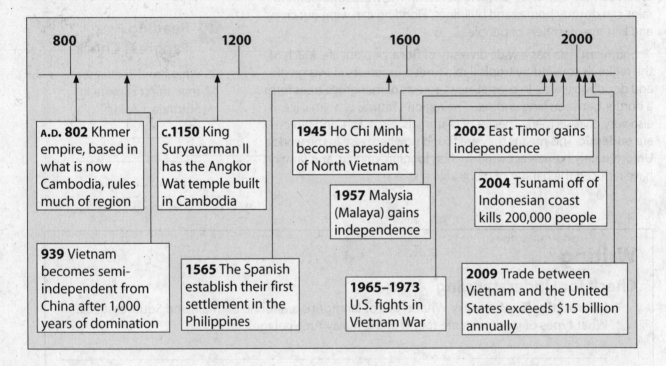

800   1200   1600   2000

A.D. 802 Khmer empire, based in what is now Cambodia, rules much of region

c.1150 King Suryavarman II has the Angkor Wat temple built in Cambodia

1945 Ho Chi Minh becomes president of North Vietnam

2002 East Timor gains independence

1957 Malysia (Malaya) gains independence

2004 Tsunami off of Indonesian coast kills 200,000 people

939 Vietnam becomes semi-independent from China after 1,000 years of domination

1565 The Spanish establish their first settlement in the Philippines

1965–1973 U.S. fights in Vietnam War

2009 Trade between Vietnam and the United States exceeds $15 billion annually

Southeast Asia

Lesson 2: History of Southeast Asia, *continued*

Kingdoms and Empires

Guiding Question *What role has trade played in Southeast Asia's history?*

Southeast Asia is known as "the Crossroads of the World" because it is located along important sea trade routes. Trade has led to the spread of many cultural influences. It has also made the region attractive to foreign powers hoping to gain riches and power through control of the trade routes.

People have lived in Southeast Asia for at least 40,000 years. For most of its history, though, the area has looked very different than it does today. During the ice age, sea levels were much lower. What is now the sea floor was once part of the mainland. The mainland was much larger, and the island area was much smaller. As the ice age waned, seas began to rise. They reached their current levels about 8,000 years ago.

People in Southeast Asia survived by hunting and gathering their food. About 6,000 years ago, people living in the fertile river valleys and deltas began growing rice. More complex societies could develop when people began settling in one place.

Eventually early metalworking societies arose in Southeast Asia. These people produced bronze tools and weapons. The most famous of these cultures was the Dong Son. This culture was centered in northern Vietnam.

By the middle of the 100s B.C., China and India had become powerful. China conquered the Red River delta and made Vietnam part of the Han empire. Vietnam remained under Chinese control for the next 1,000 years. Missionaries and traders from India travelled to Southeast Asia. These travelers spread India's culture and its religions, Hinduism and Buddhism, into Southeast Asia.

Several trade-based societies arose in the region. Funan was established in the A.D. 100s. It covered parts of what are now Cambodia, Thailand, and Vietnam. Srivijaya arose on the island of Sumatra around the A.D. 600s. This kingdom controlled the Strait of Malacca, an important trade route connecting the Pacific and Indian oceans.

Agricultural societies arose where rice could be grown. The Pagan kingdom sprang up in Myanmar's Irrawaddy delta. Vietnamese society took root in the Red River delta. The Khmer empire was centered near a large lake in Cambodia. The Khmer empire is known for architecture, especially the temple complexes Angkor Wat and Angkor Thom. These temples still stand and draw millions of visitors each year.

? Explaining

1. Why did foreign powers think they would become richer if they controlled Southeast Asia?

_____

_____

_____

✏ Marking the Text

2. Circle the name of an early metalworking society in Southeast Asia.

⚙ Contrasting

3. How did Chinese culture spread in this region? How did Indian culture spread?

_____

_____

_____

_____

? Identifying

4. Why was the Strait of Malacca important?

_____

_____

_____

Lesson 2: History of Southeast Asia, *continued*

? Summarizing

5. Describe how Islam spread through Southeast Asia.

_____

_____

_____

_____

✓ Reading Progress Check

6. What are some ways in which India influenced Southeast Asia?

_____

_____

_____

_____

_____

⚙ Determining Cause and Effect

7. Why did European traders want to find routes to Southeast Asia?

_____

_____

_____

_____

_____

Islam, the religion of Muslims, may have reached Southeast Asia as early as the A.D. 800s. It was brought by traders from the Middle East or western India who travelled the sea route to China. By the 1400s, other Islamic kingdoms were established near port cities along the main trade routes. By the 1600s, Islam was the dominant religion across most of the Malay Archipelago. The most important Islamic kingdom was located at the port of Malacca on the Malay Peninsula. Malacca grew into a powerful trading empire. Its **sultans**, or kings, ruled over most of the peninsula and the island of Sumatra.

Western Colonization

Guiding Question *How did European colonization change Southeast Asia?*

From ancient times through the Middle Ages, Chinese, Indian, and Arab traders brought precious spices to Europeans. Spices were used to flavor food, to preserve meat, and to make perfumes and medicines. Spices were in high demand but supplies were limited. This meant that traders could charge high prices. Some spices were worth more than gold.

European rulers decided they wanted to gain control of this profitable trade. During the Age of Discovery, European explorers made long sea voyages, looking for spices, as well as other riches such as gold and silver. They also wanted to spread Christianity and map the world.

In the early 1500s, Portuguese sailors discovered a way to reach India and Southeast Asia. They found they could reach these areas by sailing around the southern tip of Africa. In 1511 the Portuguese conquered Malacca. They discovered the Moluccas and the Banda Islands where cloves, nutmeg, and mace were grown. These islands became known as the Spice Islands. The wealth generated by spices made Portugal rich.

Other European powers also tried to make money from the spice trade. The explorer Ferdinand Magellan led five Spanish ships that reached the Philippines by sailing across the Pacific Ocean from Mexico. Soon the Philippines became a Spanish colony. By the middle of the 1600s, the Dutch had found a new route to the Spice Islands. They replaced the Portuguese as the area's main trading power.

In the 1800s and early 1900s, European countries gained control over other parts of Southeast Asia. Burma and Malaysia became colonies of Great Britain. Vietnam, Laos, and Cambodia became colonies of France.

Lesson 2: History of Southeast Asia, *continued*

Europeans looked for more ways to make money in Southeast Asia. They built mines and factories. They established **plantations**, or large farms on which a single crop is grown for export. Crops grown on plantations included tea, coffee, tobacco, and rubber trees. Thousands of people were brought from China and India to work in the mines and on the plantations. Many became permanent residents.

Thailand, then known as Siam, was the only nation in Southeast Asia that Europeans did not colonize. It was ruled by an **absolute monarchy** from the mid-1400s until 1932. In an absolute monarchy, one ruler has governing power over the entire country. Siam was able to remain independent by allowing free trade with European countries.

Independent Countries

Guiding Question *What events ended the colonial era in Southeast Asia?*

By the early 1900s, nearly all of Southeast Asia was controlled by foreign countries. Colonial rule was often harsh and unjust. The people sometimes violently resisted colonialism. However, countries in the region did not gain their freedom until after World War II ended in 1945.

Date	Independent Country	Colonial Power
1946	The Philippines	United States
1948	Myanmar	Britain
1949	Indonesia	Netherlands
1950s	Vietnam, Laos, and Cambodia	France
1950s	Malaysia and Singapore	Britain

The last country to gain its freedom was East Timor. It declared independence from Portugal in 1975 but was then invaded by Indonesia. Indonesia's violent rule lasted until 2002.

The newly independent countries of Southeast Asia faced many challenges. In many countries, wars and revolutions led to years of violence. In Vietnam, Communist forces defeated the French in 1954 and ruled the northern part of the country. The United States supported leaders in the south against the Communists. Fighting led to the Vietnam War, which lasted until 1975 and took more than 2 million lives. After the war ended, North Vietnam united the country under a Communist government.

✎ Marking the Text

8. In the text on the left, circle the products that are grown on plantations.

✓ Reading Progress Check

9. In what ways did European powers gain wealth from their colonies in Southeast Asia?

_____

_____

_____

✎ Marking the Text

10. Highlight countries that gained their independence from Britain in pink. Highlight countries that gained their independence from France in yellow.

⚙ Explaining

11. Why was East Timor the last country to achieve its independence?

_____

_____

_____

Marking the Text

12. In the text, underline four events that occurred in 1975.

Reading Progress Check

13. What colonial power ruled Vietnam before 1954?

_____

_____

_____

_____

In Laos, Vietnamese communists fought the government, which collapsed in 1975. In the same year, a rural communist movement called the Khmer Rouge overthrew the Cambodian government. Between 1975 and 1979, the Khmer Rouge undertook a brutal campaign of terror in which at least 1.5 million people died. In 1978, Vietnam sent soldiers to invade Cambodia and set up a new government that it controlled. This began a civil war that lasted almost 13 years.

Economic growth in China and Taiwan has helped some countries in Southeast Asia. Manufacturing is important in Thailand, Malaysia, Indonesia, the Philippines, and Vietnam. Textiles and tourism are important parts of Cambodia's economy. Singapore is one of the world's wealthiest countries.

Thailand has been a **constitutional monarchy** since 1932. In a constitutional monarchy a ruler must follow a constitution and laws. In the 1980s, Thailand adopted democratic reforms that helped the economy grow.

Myanmar has struggled since gaining independence. The military seized power and established a socialist government in 1962. Since then, the country has been closed to outside influences.

Writing

Check for Understanding

1. **Informative/Explanatory** How has location affected the development of economies in Southeast Asia?

_____

_____

_____

_____

2. **Argument** Why do you think newly independent countries in Southeast Asia had difficulties in establishing stable governments?

_____

_____

_____

Southeast Asia

Lesson 3: Life in Southeast Asia

ESSENTIAL QUESTIONS
What makes a culture unique? Why does conflict develop?

Terms to Know
primate city a city so large and important that it dominates the rest of the country
minority small ethnic group within a country
Pacific Rim the area bordering the Pacific Ocean
subsistence farming type of farming in which only enough food is grown to feed one's family
ecotourism touring natural environments such as rain forests and coral reefs

What Do You Know?
In the first column, answer the questions based on what you know before you study. After this lesson, complete the last column.

Now...		Later...
	Where do most Southeast Asian people live?	
	What religions are often practiced in Southeast Asia?	
	What are the most important crops grown in Southeast Asia?	

People and Places
Guiding Question *How is Southeast Asia's population shifting?*

Southeast Asia has fewer people than its neighbors, China and India. Compared to other areas of the world, however, Southeast Asia's population is high. The population grew quickly in the 1900s. Today, the region's growth is only slightly above the world average.

Southeast Asia is home to about 625 million people, but they are not evenly distributed throughout the region. For instance, almost 40 percent of people in the region live in Indonesia. The highest population densities are found where there is good soil and abundant water for farming. This includes coastal plains, river valleys, and deltas.

Identifying
1. In which country does about 40 percent of Southeast Asia's population live?

_____

_____

_____

networks

Southeast Asia

Lesson 3: Life in Southeast Asia, *continued*

Marking the Text

2. Read the text on the right. Highlight the definition of *primate city*.

✓ **Reading Progress Check**

3. What types of geographical areas in Southeast Asia have the highest population densities?

_____

_____

_____

⚙ **Identifying Cause and Effect**

4. Why are there so many ethnic groups in Southeast Asia?

_____

_____

_____

⇄ **Activating Prior Knowledge**

5. Why do so many people in Southeast Asia speak English, Spanish, and French?

_____

_____

_____

Since World War II, Southeast Asians have moved steadily from rural areas to cities. This movement is called urbanization. Some cities are very large. Manila, the capital of the Philippines, is home to more than 11 million people. Manila is a **primate city**. A primate city is so large and important that it dominates the rest of the country. Jakarta, the capital of Indonesia, has grown so much that it has absorbed nearby cities and become what is called a megalopolis. Today it is one of the world's largest megalopolises, with some 26 million residents.

People and Cultures

Guiding Question *How have China and India influenced Southeast Asian cultures?*

Peoples from other regions have migrated to Southeast Asia for more than 2,000 years. As a result, the region has a rich mix of peoples and cultures.

There are five main ethnic groups on the mainland, as shown in the table below. Many smaller groups, called minorities, also live in each country.

Country or countries	Main ethnic group
Myanmar	Burmese
Thailand	Siamese
Malaysia	Malay
Laos, Cambodia	Mon-Khmer
Vietnam	Vietnamese
Indonesia	Javanese
Philippines	Tagalog

People speak many languages in Southeast Asia. Most languages are native to the region. English, Spanish, French, and other European languages were brought by colonial powers.

Three-fourths of people in Southeast Asia live in rural areas. Many rural people move to the cities. Others leave to work in other countries. They send money home to help their families survive.

Education and literacy rates vary. In Vietnam, 94 percent of people can read and write. In East Timor, Laos, and Cambodia, less than 75 percent can read and write. Most schools in East Timor were destroyed during the fight for independence from Indonesia. The new government is working hard to rebuild the schools.

Southeast Asia

Lesson 3: Life in Southeast Asia, *continued*

Buddhism is the primary religion on the mainland. Islam is dominant on the southern Malay Peninsula and across Indonesia. Most people in the Philippines and East Timor are Roman Catholic. Some rural people practice animist religions. Animism is based on the belief that all natural objects, such as trees, rivers, and mountains, have spirits.

Region	Main religion
southern Malay Peninsula	
	Islam
mainland	
	Roman Catholicism
	Roman Catholicism

A wide variety of art forms reflect the region's great cultural diversity. In Thailand, plays called *ikay* feature singing, dancing, and brightly colored costumes. Indonesia, Malaysia, and Cambodia all have a rich theater tradition of shadow puppets.

Issues in Southeast Asia

Guiding Question *In what ways have human activities affected the environment in Southeast Asia?*

Much of the **Pacific Rim**, the area bordering the Pacific Ocean, has experienced rapid economic growth. This has brought great changes to some of the region's countries.

Most people earn their living by farming. Rice is a food staple and an important cash crop. Thailand and Vietnam lead the world in exports of rice. In other areas, plantations produce natural rubber, palm oil, coconuts, sugar, cacao, coffee, and spices. Many farmers grow only enough food to feed themselves and their families. This is called **subsistence farming**.

Indonesia, Malaysia, and Thailand mine large amounts of tin. Indonesia also has copper and gold deposits. Fishing is a main industry in Thailand, Indonesia, Malaysia, and the Philippines. Tourism is growing in countries such as Cambodia, Thailand, and Vietnam. **Ecotourism** allows tourists to enjoy Southeast Asia's natural environment, such as rain forests and coral reefs.

Modern technology is also a part of the economy. The tiny country of Singapore has grown into a major industrial center. Finance, communication, and information technology have also improved in Southeast Asia. Many workers are employed by European, Japanese, and U.S. companies.

✏ Marking the Text

6. Complete the table to the left on religions in Southeast Asia.

✔ Reading Progress Check

7. Which religions are most widespread in Southeast Asia?

_____
_____
_____
_____

⇄ Activating Prior Knowledge

8. Which states in our country can be considered Pacific Rim states?

_____
_____
_____

✏ Marking the Text

9. Highlight industries important in Thailand in yellow. Highlight industries important in Indonesia in blue. Highlight industries important in both countries in green.

Lesson 3: Life in Southeast Asia, *continued*

? Marking the Text

10. Highlight the environmental problems that have been caused by economic activity in Southeast Asia.

✓ Reading Progress Check

11. What is the mission of the Association of Southeast Asian Nations (ASEAN)?

_____

_____

_____

In 1968 Indonesia, Malaysia, the Philippines, Singapore, and Thailand formed the Association of Southeast Asian nations (ASEAN). The organization tries to increase economic development, social progress, and cultural development in the region. It promotes peace and security across the region.

Southeast Asia faces many challenges. Cambodia, Laos, Myanmar, and Vietnam are among the poorest countries in the world. In many countries, the gap between the rich and poor is getting bigger. The rapid growth of cities like Manila and Jakarta has led to overcrowding and water shortages.

The environment has also suffered. Tin mining has created wastelands in some areas. Tropical forests are disappearing due to commercial logging and farming. Dams along the Mekong River generate electricity but also harm the fishing industry. Vietnam, the Philippines, and Malaysia are arguing over claims to oil and natural gas reserves in the South China Sea.

In some areas, economic and social progress has been slowed by political problems. This is particularly true in Myanmar. The country became isolated from the rest of the world after the military took over the government in 1962. The brutal government refused to allow democratic elections or political protest. In 2011 Myanmar began to allow more freedoms. In 2012 the United States reestablished diplomatic ties with the country.

Writing

Check for Understanding

1. **Informative/Explanatory** What kinds of problems have been caused by the rapid urbanization in Southeast Asia?

_____

_____

2. **Narrative** Write a postcard to a friend describing what you might see on a tour of the farm areas of rural Southeast Asia.

_____

_____

_____

networks

South Asia

Lesson 1: Physical Geography of South Asia

ESSENTIAL QUESTION
How does geography influence the way people live?

Terms to Know
subcontinent a large landmass that is part of a continent
alluvial plain an area built up by rich, fertile soil left by river floods
delta an area where soil is dropped at the mouth of a river
atoll a small, ring-shaped island made of coral
monsoon seasonal wind patterns that bring heavy rains for several months of the year
cyclone a storm with high winds and heavy rains

Where in the World: South Asia

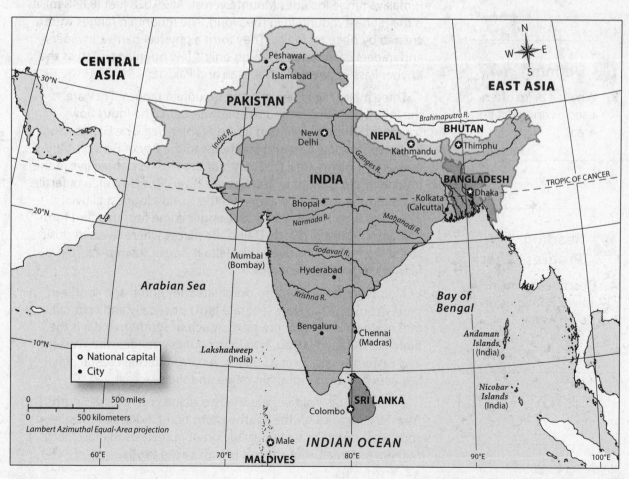

networks

South Asia

Lesson 1: Physical Geography of South Asia, *continued*

Marking the Text

1. Read the text on the right. Highlight the names of the seven countries that make up South Asia.

Analyzing

2. Why was the Khyber Pass so important to South Asia for centuries?

_____

_____

_____

Defining

3. Why is a *delta* often used as an agricultural area?

_____

_____

_____

Reading Progress Check

4. Describe the main physical regions of South Asia.

_____

_____

_____

_____

_____

_____

South Asia's Physical Features

Guiding Question *What physical features make South Asia unique?*

South Asia is a **subcontinent**, a geographically unique part of a larger continent. India is the largest country in the region of South Asia. The other six countries are Pakistan, Nepal, Bhutan, Bangladesh, Maldives, and Sri Lanka.

Geography of South Asia	
Northern	mountains and plains
Central	mountains and rivers
South	Deccan Plateau
Islands	Sri Lanka and Maldives

Three mountain ranges form South Asia's northern border. They are the Hindu Kush, the Karakoram, and the Himalaya. The Himalaya range includes Mount Everest. At 29,028 feet (8,848 m), it is the highest mountain in the world. The mountain ranges were created by plate tectonics. They form a physical barrier. Invaders and traders could enter through only a few openings, such as the Khyber Pass between Afghanistan and Pakistan.

Three major rivers begin in the mountain ranges. They are the Indus, the Ganges, and the Brahmaputra. The Indus flows southward through Pakistan to the Arabian Sea. The Ganges and Brahmaputra flow east and southeast to the Bay of Bengal. They cross a vast plains area. One-tenth of the world's people live in the **alluvial plain** created by the Ganges River. This is an area of fertile soil deposited by floodwaters. It is the world's longest alluvial plain. The Brahmaputra and the Ganges come together and form the largest **delta** on Earth. Deltas are places where rivers deposit soil at the mouth of a river. This delta is one of the most fertile farming regions in the world.

Mountains and rivers also dominate the central and southern parts of South Asia. They separate India physically and culturally into northern and southern parts. Much of southern India is the high, flat Deccan Plateau. Two low mountain ranges, the Western and Eastern Ghats, form its edges. A narrow, fertile coastal plain lies between each mountain range and the seacoast.

Sri Lanka and Maldives are the two island countries of South Asia. Sri Lanka lies off the southeastern tip of India. Maldives lies southwest of India's tip. It is made up of numerous islands. Many of them are small, ring-shaped islands called **atolls**.

Lesson 1: Physical Geography of South Asia, *continued*

South Asia's Climates

Guiding Question *How does climate affect people's lives in South Asia?*

About half of South Asia has a tropical climate. Much of the northern half enjoys a warm, temperate climate. There are cool highlands in the north and scorching deserts to the west. Much of the region's climate is a result of seasonal wind patterns called **monsoons**. Little or no rain falls for eight months of the year. In May and early June, temperatures soar, causing the wind direction to change. Winds from the Indian Ocean bring heavy rains and flooding. Without the monsoons, the region could not grow enough food for its people. However, the floods also damage property and cause loss of life.

Large, swirling storms called **cyclones** often slam into the coast along the Bay of Bengal. The strong winds push water onto the shore, flooding low-lying areas far inland. A cyclone can kill tens of thousands of people. River deltas are especially vulnerable.

Much of South Asia has a tropical wet/dry climate. As a result of monsoon wind patterns, there are just three seasons: hot, cool, and wet. The hot and cool seasons are dry. Tropical wet climates get plenty of rain all year. They can be found along the western coast of India, southern Sri Lanka, and the Ganges delta in Bangladesh. They have thick, green vegetation.

Not all of South Asia is drenched by monsoons. Parts of the Deccan Plateau get little rain because the Western Ghats block the winds and rains of the monsoons. The region's driest climate is in the northwest. The Thar Desert lies between Pakistan and India. Vegetation is mostly low, thorny trees and dry grasses.

The mountains to the north affect the climate of lower areas. In winter, the Himalaya block cold winds sweeping down from Central Asia. This forms a temperate zone across Nepal, Bhutan, northern Bangladesh, and northeastern India. Farther south, the elevation of the Deccan Plateau creates another temperate climate area.

South Asia's Natural Resources

Guiding Question *Which natural resources are most important to South Asia's large population?*

South Asians rely on rivers for irrigation, drinking and household water, and transportation. Fast-flowing rivers can also be used to generate electricity. Rivers are considered sacred in Hinduism, the main religion in India.

 Determining Central Ideas

5. What might happen in South Asia if there were no monsoons?

_____

_____

_____

_____

Comparing

6. How does a cyclone differ from a monsoon?

_____

_____

_____

 Marking the Text

7. Highlight the text that tells how the Himalaya affect the climate.

✓ **Reading Progress Check**

8. What positive and negative effects do the monsoons have on the lives of people in South Asia?

_____

_____

_____

_____

South Asia

Lesson 1: Physical Geography of South Asia, *continued*

India has most of South Asia's mineral resources. Nepal and Sri Lanka also have mineral resources.

South Asia's Mineral Resources	
India	iron ore, manganese, chromite, mica
Nepal	mica, copper
Sri Lanka	sapphires, rubies, graphite

Iron ore, manganese, and chromite are used to make steel. Mica is a rock used to manufacture electrical equipment. Graphite is the "lead" used in pencils. It is also used in batteries and as a lubricant.

There are important petroleum reserves in northern Pakistan and near the Ganges Delta, as well as in the Arabian Sea west of Mumbai. Overall, though, South Asia depends on imported oil. There are natural gas fields in southern Pakistan and in Bangladesh. India has uranium in the Eastern Ghats. This uranium is used in the country's nuclear power plants.

Valuable timber resources include teak, sal, and sandalwood. However, forests are more than just resources for wood products. They take in carbon dioxide, a greenhouse gas, and release oxygen. Tree roots hold soil in place, reducing erosion. Indian forests are home to three of Earth's most endangered species: the tiger, the Asian elephant, and the one-horned rhinoceros. South Asians are working to reverse some of the region's wildlife losses. The creation of wildlife reserves and laws controlling hunting and logging have started to make a difference.

⚙ Analyzing

9. What might be the consequences of cutting down a forest in South Asia?

_____

_____

_____

☑ Reading Progress Check

10. Think about how people use rivers in South Asia. How is it similar to how rivers are used in other parts of the world?

_____

_____

_____

_____

Writing

Check for Understanding

1. Informative/Explanatory What physical feature makes South Asia unique? How?

_____

_____

2. Argument What feature of South Asia's climate has the most significant effect on the region? Explain

_____

_____

South Asia

Lesson 2: History of South Asia

ESSENTIAL QUESTION
How do governments change?

Terms to Know

caste social class into which a person is born and cannot change
reincarnation the belief that after a person dies, his or her soul is reborn into another body
Raj the period of time in which Great Britain controlled India as a part of the British Empire
boycott to refuse to buy items from a particular country or company
civil disobedience the use of nonviolent protests to challenge a government or its laws
nuclear proliferation the spread of powerful nuclear weapons among nations

When did it happen?

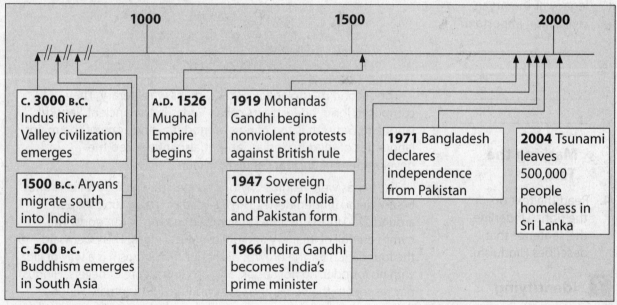

c. 3000 B.C. Indus River Valley civilization emerges

1500 B.C. Aryans migrate south into India

c. 500 B.C. Buddhism emerges in South Asia

A.D. 1526 Mughal Empire begins

1919 Mohandas Gandhi begins nonviolent protests against British rule

1947 Sovereign countries of India and Pakistan form

1966 Indira Gandhi becomes India's prime minister

1971 Bangladesh declares independence from Pakistan

2004 Tsunami leaves 500,000 people homeless in Sri Lanka

Early South Asia

Guiding Question *How did South Asia's early history lay the foundation for modern life in the region?*

One of the oldest known civilizations was located near the Indus River. Called the Indus Valley civilization, it dates back to 3500 B.C. It formed at about the same time as other river-valley civilizations around the world. Archaeologists have discovered two large cities and dozens of smaller settlements. The people of this culture left written records. Most were farmers, but crafts also flourished.

Describing

1. What was the Indus Valley civilization like?

_____

_____

_____

South Asia

Lesson 2: History of South Asia, *continued*

 Defining

2. How did the *caste* system affect life in South Asia?

_____

_____

_____

_____

Explaining

3. Why was the literary legacy of the Aryan civilization important?

_____

_____

_____

_____

Marking the Text

4. Read the text on the right. Underline the sentence that describes Hinduism.

Identifying

5. Who was the Buddha?

_____

_____

_____

_____

Evidence shows that people traded over long distances. The Indus Valley culture lasted about 1,000 years. No one knows why it ended. Natural disasters or enemy invasions are possible causes.

About 1500 B.C., the Aryans swept into what is now India. They probably came from Russia or central Asia. Once in India, they settled down to become farmers. The Aryan civilization lasted about 1,000 years. It left behind two important legacies.

One legacy was the caste system, shown below. **Castes** were social classes. At the top were priests, called Brahmans. At the bottom were laborers and peasants. The caste system had a deep impact for thousands of years and caused great inequality. People born into a lower class could not move up in society. Only after India won independence in 1947 was the caste system outlawed.

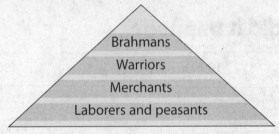

The second major legacy of the Aryans was literary. They composed long, poetic texts called Vedas in the ancient Sanskrit language. Sanskrit is the parent of modern Hindi, one of the major languages of modern India. Sanskrit also influenced the development of ancient Greek and Latin.

The Vedas were religious hymns handed down orally for centuries before being written down. The *Rig Veda* probably took shape around 1200 B.C. It is a series of hymns honoring Aryan gods. The hymns are full of imagery and philosophical ideas. This poem laid the foundation for the growth of Hinduism. Hinduism is a way of life with no founder, no holy book, and no central core of beliefs. Hindus pay respect to the Vedas and take part in religious rituals. Hindus believe in **reincarnation**, or the rebirth of the soul in another body.

Around 500 B.C., two new religions arose in response to Hinduism and the caste system. One was Jainism, based on the Hindu principle of noninjury. Jains turned from farming to trade so they would not have to kill or injure any living creature.

The other new religion was founded by a prince named Siddhārtha Gautama. He gave up his wealthy lifestyle and traveled in poverty, seeking spiritual truth. When he reached his goal, he became known as "the Buddha," or "the enlightened one." He passed on what he believed to be the Four Noble Truths.

South Asia

Lesson 2: History of South Asia, continued

1 Life is full of suffering.

2 Suffering is caused by selfish desire.

3 Conquering desire can stop suffering.

4 Desire can be conquered by the Eightfold Path: right view, right intention, right speech, right action, right way of living, right effort, right mindfulness, and right concentration.

Like the Jains, Buddhists largely rejected the caste system. Today, Hinduism remains the major religion of India. Buddhism has spread to other Asian countries, while Jainism remains a minor religion in India.

Around 200 years after the Aryan civilization faded, the Mauryas conquered much of South Asia. Their most famous ruler was Ashoka. He was a warrior who turned to nonviolence after converting to Buddhism around 260 B.C. His conversion influenced many people. Trade and culture thrived under his rule.

Hundreds of years later, the Gupta Empire unified much of northern India. Under Chandragupta I, science, medicine, mathematics, and the arts flourished. Gupta scholars developed the decimal system in mathematics that we still use today.

During the 1500s and 1600s, the Mughal Empire flowered in India. The Mughals were Muslim and the first rulers to be members of a minority religion. During this era, many South Asians converted to Islam. Some of the Mughals were tolerant. Akbar the Great encouraged freedom of religion. Culture, science, and the arts flourished under the Mughals. A Mughal emperor, Shah Jahan, built the Taj Mahal, an architectural monument in memory of his beloved wife.

Modern South Asia

Guiding Question *How has conflict in South Asia led to change?*

Beginning in the 1600s, British traders established settlements in India. As the Mughal Empire declined, the traders became more powerful in the region. The British were especially interested in textiles, timber, and tea. After a bloody rebellion in 1857, the British government took direct control of most of South Asia.

Comparing and Contrasting

6. How are the religions of Jainism and Buddhism alike? How are they different?

_____

_____

_____

_____

_____

Activating Prior Knowledge

7. What is the decimal system?

_____

_____

Marking the Text

8. Read the text on the left. Highlight the names of South Asian civilizations or empires.

Reading Progress Check

9. What two important legacies did the Aryans leave behind?

_____

_____

_____

networks

South Asia

Lesson 2: History of South Asia, *continued*

Marking the Text

10. Highlight the names of two leaders of the Indian independence movement.

Reading Progress Check

11. How did Mohandas K. Gandhi and Jawaharlal Nehru approach the challenge of gaining independence from British rule?

_____

_____

_____

_____

India became a British colony. Although the British built railways, schools, and ports, the Indians resented a foreign presence in their land. In the late 1800s, an independence movement began.

In 1885 Indian supporters of independence formed the Indian National Congress. But the British did not want to give up the **Raj**, as their imperialist rule of India was called. The National Congress responded by encouraging a **boycott**, refusing to buy or use imported British goods.

In the early 1900s, two members of the Congress became leaders. Mohandas K. Gandhi was opposed to violence. His most powerful weapon was **civil disobedience**, or nonviolent resistance, to British rule. He was joined by a younger leader, Jawaharlal Nehru. Together, they finally persuaded the British to leave South Asia in 1947.

South Asia has been troubled by religious and cultural divisions between Hindus and Muslims. As part of the independence settlement, the subcontinent was split into two countries. India was mainly Hindu. Pakistan was mainly Muslim. Tensions between the two countries developed. They have fought several wars and are involved in a dispute over the Kashmir region. In the late 1990s, both countries developed nuclear weapons. This **nuclear proliferation**, or spread of powerful atomic weapons, could make conflict between the two nations dangerous.

Writing

Check for Understanding

1. Informative/Explanatory Who were the Guptas, and what did they accomplish?

_____

_____

_____

2. Informative/Explanatory Why are Hindu-Muslim conflicts in South Asia so significant to the history of the region?

_____

_____

_____

South Asia

Lesson 3: Life in South Asia

ESSENTIAL QUESTIONS

What makes a culture unique?

Terms to Know

sitar long-necked, stringed instrument
green revolution the effort to increase crop yields using irrigation, fertilizers, and high-yielding crops
cottage industry small business that employs people in their homes
ecotourism a type of tourism in which people visit an area to enjoy its natural wonders
outsourcing hiring companies or workers in other countries to do work
dalit the oppressed "untouchables" of Indian society

What Do You Know?

In the first column, answer the questions based on what you know before you study. After this lesson, complete the last column.

Now...		Later...
	Where do most people live in South Asia?	
	What languages do people in South Asia speak?	
	What religions do people in South Asia practice?	
	What challenges face the people of South Asia?	

People and Places

Guiding Question *What are the major population patterns of South Asia?*

South Asia is about half the size of the lower 48 states of the United States, but more than 1.5 billion people live there. That is about five times the American population. The largest countries, in area and population, are India, Pakistan, and Bangladesh. India is the world's second-most-populous country, after China. However, the birthrate in India is higher than in China. This means that by 2030, India will be the world's most populous country.

Marking the Text

1. Read the text on the left. Highlight the names of the three largest countries in South Asia.

networks

Lesson 3: Life in South Asia, *continued*

Determining Central Ideas

2. Why might population growth in South Asia be a major problem in the years ahead?

_____

_____

_____

_____

Reading Progress Check

3. Why is India projected to overtake China as the most populous country by 2030?

_____

_____

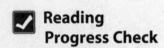

Comparing

4. What is one way in which South Asia and the United States are similar?

_____

_____

Marking the Text

5. Read the text on the right. Highlight the names of the languages spoken in Pakistan, Bangladesh, and Sri Lanka.

Population Profile of India

- Population: 1.22 billion
- Percentage of the world's people: 17.3%
- Median age: 25 years

South Asians live mainly in areas that are good for farming. A high percentage of Indians live in the Ganges Plain. Agriculture plays a major role in India's economy. About 7 of every 10 people live in small villages. However, more and more Indians leave their villages every year. Indian cities are experiencing rapid growth.

The country's business center, Mumbai, has about 20 million people. It is India's largest city and the fourth-largest city in the world. Population *density* is the number of people living in a given unit of area. In Mumbai, the population density is 80,100 per square mile (30,900 per sq. km), seven times the world's average.

The rapid growth of South Asia has put a strain on its resources. Cities struggle to provide essential services, and air and water pollution have increased with overcrowding.

People and Cultures

Guiding Question How are the diverse cultures in South Asia rooted in ethnic and religious traditions?

The cultures of South Asia are highly diverse. We think of the United States as a "melting pot." The same might be said of South Asia.

India has two official languages, Hindi and English. However, 15 other languages are each spoken by millions of people. After India won independence in 1947, boundaries of Indian states were based mainly on ethnic groups and languages. Indians are proud of their ethnic and language heritage.

The most common languages in Pakistan are Urdu and English. Urdu, like Hindi, developed from ancient Sanskrit. In Bangladesh, people speak Bangla, a version of Indian Bengali. Most people in Sri Lanka speak Sinhalese, which has its own a!phabet.

Two great epic poems of ancient India, the *Ramayana* and the *Mahabharata*, embody Hindu social and religious values. South Asian children of all religions know the plots and characters of these epics. The region is also a center for classical dance. Music is a thriving art. Ravi Shankar is probably India's best-known musician. He plays the **sitar**, a stringed instrument. Motion pictures arrived in India in 1896. The country now has the world's largest film industry.

Lesson 3: Life in South Asia, *continued*

Six main religions are practiced in South Asia. Sikhism developed in the 1500s, nearly 2,000 years after Buddhism and Jainism. It was also a reaction to Hinduism. Sikhs reject the Hindu caste system. Like Muslims and Christians, they are monotheists: they believe in only one God.

Religions of South Asia		
• Hinduism	• Islam	• Buddhism
• Jainism	• Sikhism	• Christianity

Life in South Asia centers on the family. Often, several generations of family members live together. Within the family, age and gender play important roles. Elders are respected, and females are often subordinate to males. Arranged marriages are still common, but less so than in the past. Parents often introduce couples to each other but allow them to decide for themselves whether to marry. It is still important for couples to get approval to marry from their parents.

A typical South Asian meal would include rice, legumes, and flatbreads. A combination of spices called curry flavors many dishes. People in South Asia do not eat much meat, partly because of religious guidelines.

Issues in South Asia

Guiding Question *What impact do economic and environmental issues have on life in South Asia?*

Many South Asians are farmers, but good cropland is scarce. How do South Asian farmers grow enough food for the huge population? The **green revolution** has helped increase crop yields. It involves the use of irrigation, fertilizers, and high-yielding crops. With these improvements, India has not had to import food since the 1970s. This situation may change, though. These methods are no longer increasing productivity enough to meet the growing need for food.

Some South Asians make a living from mining or fishing. Others own or work for **cottage industries**. These are small businesses that employ people in their homes. They include textile weaving, making jewelry and furniture, and wood carving. Advanced technology is a fast-growing part of the economy. Indian computer specialists, engineers, and software designers are in demand throughout the world. Another part of the economy is **ecotourism**, which combines recreational travel and environmental awareness.

Comparing

6. How is Sikhism similar to other religions in South Asia?

_____

_____

_____

Reading Progress Check

7. How do South Asian culture and American culture compare?

_____

_____

_____

_____

Defining

8. Why has the *green revolution* been important for South Asians?

_____

_____

_____

_____

Defining

9. What is the definition of a *cottage industry*?

_____

_____

_____

South Asia

Lesson 3: Life in South Asia, *continued*

Marking the Text

10. Read the text on the right. Underline the sentences that describe the situation of *dalits* in India.

✔ Reading Progress Check

11. Why might a U.S. company outsource its customer-service operations to a South Asian company?

_____

_____

_____

_____

Because English is so widely spoken in India, the country is well suited for **outsourcing**. This occurs when a company hires another company or an individual outside their home territory to do work. Companies in the United States often outsource to other countries where workers are more flexible or willing to work for lower wages.

South Asia faces many challenges. One is the long-standing conflict between India and Pakistan over Kashmir. Two of India's prime ministers have been assassinated in recent decades. Both were killed by extremists who opposed their views. Another issue in India is the oppression of ***dalits***, the so-called "untouchables." For centuries, they have been discriminated against as outcasts.

In Sri Lanka, a civil war between Buddhists and Hindus lasted several decades and cost more than 60,000 lives. In Nepal, a rebel group has been trying since 1996 to overthrow the democratic government. Tribal people control large parts of the Pakistan-Afghanistan border.

Rapid growth and lack of infrastructure have led to air and water pollution in many parts of the region. Deforestation is also a problem. Most of India's forests have been cut down. Water quality, wildlife habitats, and climate have been affected as a result.

South Asia has had some success in meeting these challenges. The war in Sri Lanka is over. India has taken strict measures to control air and water pollution. Wildlife conservation efforts have increased as well.

Writing
Check for Understanding

1. Informative/Explanatory Why is South Asia sometimes called a religious and ethnic "melting pot"?

_____

_____

_____

2. Informative/Explanatory What are three internal conflicts South Asian countries have faced?

_____

_____

_____

netw⚬rks

Central Asia, the Caucasus, and Siberian Russia

Lesson 1: Physical Geography of the Regions

ESSENTIAL QUESTION
How does geography influence the way people live?

Terms to Know
tundra treeless zone found near the Arctic Circle or at high mountain elevations
taiga area of large coniferous forests
permafrost layer of permanently frozen ground beneath the surface soil and rocks
steppe partly dry grassland area
deciduous trees that lose their leaves in the fall

Where in the World: Central Asia, the Caucasus, and Siberian Russia

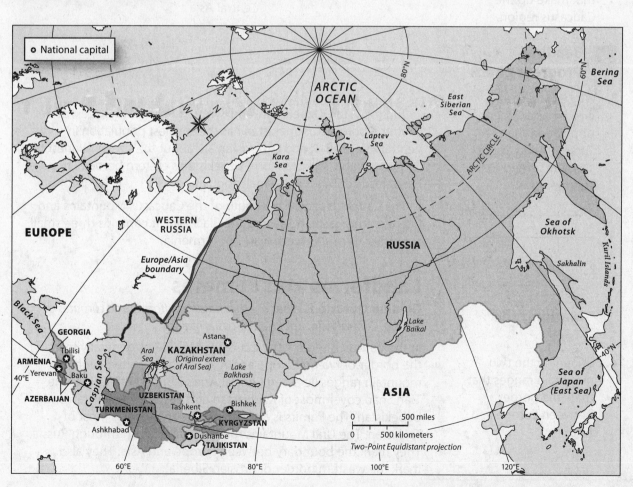

networks

Central Asia, the Caucasus, and Siberian Russia

Lesson 1: Physical Geography of the Regions, *continued*

? Listing

1. List the countries and features that border Kazakhstan.

_____

_____

_____

_____

✏ Marking the Text

2. Circle the countries that make up the Caucasus region.

✔ Reading Progress Check

3. Why is Kazakhstan, which has a larger population than Turkmenistan, more sparsely settled than Turkmenistan?

_____

_____

_____

_____

✏ Marking the Text

4. Underline the two mountain ranges that mark the border between Europe and Asia.

Regions

Guiding Question *Which countries make up the regions?*

Siberia is in the eastern part of Russia. It stretches from the Ural Mountains in the west to the Pacific Ocean in the east. Siberia is 25 percent larger than Canada, the world's second-largest country.

Central Asia is made up of five countries, as shown below. Kazakhstan is bordered by China on the east, the Caspian Sea on the west, and Russia to its north. Kazakhstan is the largest country in Central Asia, and it is also the most sparsely populated.

Turkmenistan, Uzbekistan, and Kyrgyzstan lie along Kazakhstan's southern border. Turkmenistan has the smallest population in Central Asia. Both it and Uzbekistan are about the size of Minnesota. Kyrgyzstan and Tajikistan form Uzbekistan's eastern border. They are both small countries with small populations.

The Caucasus region lies south of the Caucasus Mountains and between the Caspian Sea and the Black Sea. It contains three small countries: Georgia, Azerbaijan, and Armenia.

Landforms and Climates

Guiding Question *What are the major landforms and climates of Siberia, Central Asia, and the Caucasus region?*

In the Caucasus region, the Caucasus Mountains generally mark the border between Europe and Asia. To the east, two high mountain ranges lie along Central Asia's border with China. The Tian Shan cover most of Kyrgyzstan and eastern Kazakhstan and Uzbekistan. The Pamirs is a smaller range that covers most of Tajikistan. The Ural Mountains run north and south through Russia. They mark the boundary between Europe and Asia. They also mark the western border of Russian Siberia.

These regions also have plains and deserts. In Siberia, the west Siberian plain extends from the Ural Mountains east to the Yenisey River. East of the Yenisey River, the land rises to form the central Siberian plateau.

Northern Siberia is mostly **tundra**, a treeless zone near the Arctic Circle or at high mountain elevations. South of the tundra is the **taiga**, a zone of coniferous forest. The taiga is swampy because, like the tundra, it is covered in **permafrost**. This is a layer of permanently frozen ground that lies beneath the surface of soil and rocks. Southern portions of the western plains and central plateau contain dry grasslands called **steppe**.

Most of Kazakhstan has dry grassland plains and plateaus, with lowlands on the coast along the Caspian Sea. The Kyzl Kum desert stretches from eastern Kazakhstan to Uzbekistan. The Kara-Kum desert covers most of Turkmenistan.

Much of Central Asia has an arid climate. The steppe of southwest Siberia and northern Kazakhstan has a semiarid climate. Moving north and east from there, summers become shorter and cooler, and the winters become colder.

Mountain areas of Central Asia have a humid continental climate. Mountain valleys have hot, dry summers and cold winters. Mountain foothills are cooler and get more rain.

The Caucasus region is similar to eastern Central Asia. However, the Caspian and Black seas make the region's summers cooler and winters warmer. The Black Sea gives Georgia's coastal lowlands a humid subtropical climate.

The regions have a number of important waterways.

Waterway	Location
Caspian Sea	saltwater lake; separates the Caucasus and Central Asia
Aral Sea	saltwater lake; straddles the Kazakhstan-Uzbekistan border
Lake Baikal	in southeastern Siberia; world's largest freshwater lake by volume
Syr Dary'ya and Amu Dar'ya	rivers in Central Asia; flow from mountains across deserts
Ob', Irtysh, Yenisy, Lena	rivers in Siberia; flow north to Arctic Ocean
Amur	river in Siberia; flows eastward; part of it forms border between Russia and China

✎ Marking the Text

5. Highlight the text that defines each of the following: *tundra*, *taiga*, and *steppe*.

❓ Describing

6. Describe how the Caspian and Black seas affect the climate of the Caucasus region.

_____

_____

_____

_____

❓ Identifying

7. Which waterway is the world's largest freshwater lake by volume?

_____

_____

_____

✔ Reading Progress Check

8. What type of terrain is common in Siberia, Central Asia, and the Caucasus region?

_____

_____

_____

networks

Central Asia, the Caucasus, and Siberian Russia

Lesson 1: Physical Geography of the Regions, *continued*

Determining Word Meaning

9. Why does Central Asia have few trees?

_____

_____

Marking the Text

10. Highlight areas that have coal resources.

Reading Progress Check

11. Which three resources are economically important in all three regions?

_____

_____

_____

Natural Resources

Guiding Question *Which natural resources of Siberia, Central Asia, and the Caucasus are economically important?*

Siberia's vast taiga contains about 20 percent of all the world's trees. The area's economic value is limited, however. There are few roads, and the quality of the larch wood that grows there is poor. Central Asia has few trees because of its arid climate. In some Caucasus lowlands, oak and other **deciduous** trees grow. Deciduous trees lose their leaves in the fall. In the mountains are pine, fir, and other coniferous trees.

Siberia's Ob' River basin has oil and natural gas fields. The Central Siberian Plateau supplies most of Russia's coal. Eastern Siberia also holds most of Russia's gold, lead, zinc, and iron ore. The tundra region near the mouth of the Yenisey River is one of the world's largest producers of nickel and platinum.

Kazakhstan, Uzbekistan, and Turkmenistan also have oil and gas resources. Large coal deposits are found in Kazakhstan, Tajikistan, and Uzbekistan. Kazakhstan produces uranium. The mountains of Tajikistan and Kyrgyzstan have rich mineral resources, as do the eastern mountain areas of Kazakhstan and Uzbekistan. Gold, mercury, copper, iron, tin, lead, zinc, and other metals are mined there.

Most of the Central Asian countries use their rivers to produce electricity. This is also true of the Caucasus region. Only Azerbaijan is a major oil and gas producer in this region.

Writing

Check for Understanding

1. Informative/Explanatory Why is the economic value of Siberia's taiga limited?

_____

_____

2. Informative/Explanatory How do you think the cold climate in much of the region affects the way people there make a living?

_____

_____

Central Asia, the Caucasus, and Siberian Russia

Lesson 2: History of the Regions

ESSENTIAL QUESTION
How did the settlement of Siberia progress over time?

Terms to Know
pastoral a way of life based on herding animals
collective a large, government-run farm
irrigate to bring a supply of water to a dry area, often for farming

When did it happen?

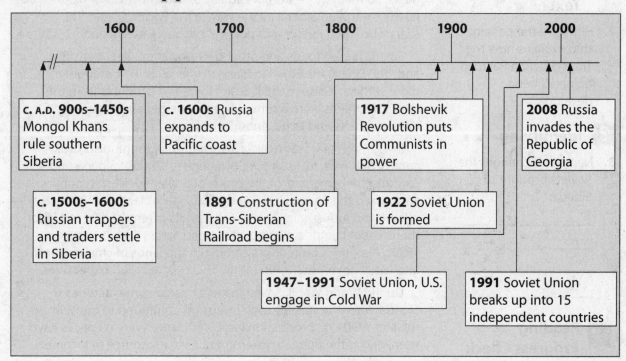

c. A.D. 900s–1450s Mongol Khans rule southern Siberia

c. 1500s–1600s Russian trappers and traders settle in Siberia

c. 1600s Russia expands to Pacific coast

1891 Construction of Trans-Siberian Railroad begins

1917 Bolshevik Revolution puts Communists in power

1922 Soviet Union is formed

1947–1991 Soviet Union, U.S. engage in Cold War

1991 Soviet Union breaks up into 15 independent countries

2008 Russia invades the Republic of Georgia

Central Asia, the Caucasus, and Siberian Russia

Lesson 2: History of the Regions, *continued*

 Defining

1. Use the word *pastoral* in a sentence.

   _____

   _____

   _____

   _____

✎ **Marking the Text**

2. Highlight the passage that explains how the Trans-Siberian Railroad helped Siberia develop.

❓ **Identifying**

3. Name two changes the Soviets brought to Siberia.

   _____

   _____

   _____

   _____

☑ **Reading Progress Check**

4. How did control of Siberia change over time?

   _____

   _____

   _____

   _____

Siberia

Guiding Question *How did the settlement of Siberia progress over time?*

Sometime after 1000 B.C., Turkic, Iranian, Mongol, and Chinese people began migrating into southern Siberia. These early people were nomads. Some were hunter-gatherers. Others were **pastoral**, which means their lives were based on herding animals. In the 200s B.C., invaders came from Manchuria in northeast China. They drove many groups onto the central Siberian plateau. Invasions by the Manchurians, Huns, Mongols, and Tartars followed.

Russian traders moved east and built small forts and trading posts. Some of these sites eventually became towns. By 1700, Russia controlled Siberia all the way to the Pacific Ocean. The czars who ruled Russia sent political prisoners to Siberia.

Siberia lacked roads and other transportation. Between 1891 and 1905, the Trans-Siberian Railroad was built. This brought in more settlers. It also made it easier to export products from the area. Coal mines were opened. The steppe began producing dairy products and large amounts of grain.

In 1917, the czar was overthrown, and the Communists took control of Russia. In 1922, the Communists brought Siberia under control. It became part of the Union of Soviet Socialist Republics (also called the Soviet Union and USSR). Soviet leaders increased mining and industry in Siberia. The Soviets often used forced labor there. Labor camps called gulags spread across Siberia in the 1930s. Also, the Communists combined the lands of small farmers into large, government-run farms. These were called **collectives**.

During World War II, Soviet leaders moved some factories to Siberia. After the war, Siberian industries continued to grow. In the late 1990s, oil production increased. Since then, oil prices have risen, giving the Russian government a steady source of income.

Central Asia

Guiding Question *How have invaders affected Central Asia's history and culture?*

For most of their history, the countries of Central Asia have been part of other empires. These include the Chinese, Huns, and others. Many conquerors wanted to control the Silk Road, the network of trade routes that linked China and Europe. Nomad groups that could not defeat the invaders often retreated into the region's deserts and dry plains. They were able to live in freedom there.

Central Asia, the Caucasus, and Siberian Russia

Lesson 2: History of the Regions, *continued*

In the A.D. 700s, the Arabs brought the religion of Islam to the region. The Arabs were replaced by other conquerors.

Arabs ⟩ Persians ⟩ Mongols

In the mid-1300s, a conqueror named Timur overthrew the Mongols. Timur made the Silk Road city of Samarqand the capital of his empire. It became a center of culture and learning. Islamic centers of learning called madrassas were created there.

In the 1700s, Czar Peter the Great expanded Russia into parts of Central Asia. Russians wanted to grow cotton in Kazakhstan and Uzbekistan. To increase production, they began to **irrigate,** or bring water to, the land. They also expanded railroads.

When the Russian Revolution began in 1917, parts of Central Asia declared their independence. Five republics were formed in the region. But by 1920, Communist forces had regained control. The republics became part of the Soviet Union.

Soviet rule brought positive changes. They built dams on rivers to generate electricity. The dams also provided water for irrigation. The Soviets also increased industry by opening mines and factories to develop the region's mineral wealth.

There were also negative changes. The Soviets resettled large numbers of Russians and other Europeans in the area. Local farmers and nomadic herders were forced into collectives. The Soviets tried to uproot local cultures. They closed mosques to end the practice of Islam in the region.

When the Soviet Union collapsed in 1991, the republics declared their independence. The region's present-day countries were formed. Today, most struggle with a number of problems, such as modernizing their economies. They also must deal with ethnic tensions and dictatorial rulers.

The Caucasus

Guiding Question *How did the countries of Georgia, Armenia, and Azerbaijan develop?*

The history of conquest and migration in the Caucasus region makes it one of the most ethnically complex places in the world. In ancient times, Georgia and Armenia were both independent before they came under the control of powerful neighbors.

✏ Marking the Text

5. Underline the passage that explains why the Russians wanted to irrigate parts of Central Asia.

Identifying

6. List four negative changes the Soviets brought to Central Asia.

_____

_____

_____

_____

_____

_____

_____

☑ Reading Progress Check

7. Why was Central Asia difficult to conquer and control completely?

_____

_____

_____

_____

_____

Central Asia, the Caucasus, and Siberian Russia

Lesson 2: History of the Regions, *continued*

Marking the Text

8. Circle all the different rulers of the Caucasus from A.D. 300 until World War I.

Reading Progress Check

9. Which Caucasus country did not have a history of independence before the end of Soviet rule?

_____
_____
_____
_____

Around A.D. 300, each country threw off Persian control. They established their own kingdoms again. They converted to Christianity. In the A.D. 600s, Islam came to the region.

The Mongols conquered much of the region in the 1200s. It became part of Timur's empire in the 1300s. Then in the 1400s, the Persians and Ottoman Turks competed for control of the area. Christian Armenians and Georgians suffered under Persian and Turkish control and asked Russia for protection. By World War I, Russia controlled the area.

After the Russian Revolution, Georgia, Armenia, and Azerbaijan briefly formed independent states. The Soviet Union took them over in 1922 and changed the area from agricultural to an urban industrial region.

During World War II, Germany invaded the Caucasus. Some ethnic groups were accused of helping the Germans. After the war, these groups were sent to other republics in the USSR. The Soviets punished other Caucasus people for loyalty to their ethnic identity and culture.

When the Soviet Union dissolved in 1991, Georgia, Armenia, and Azerbaijan declared their independence. Since independence they have struggled with economic changes, ethnic tensions, and border conflicts.

Writing

Check for Understanding

1. **Informative/Explanatory** How did Soviet collectives change agriculture in Siberia and Central Asia?

_____
_____
_____

2. **Argument** Do you believe that the Soviet Union's control in Central Asia was beneficial to the people of the region? Explain.

_____
_____
_____
_____

networks

Central Asia, the Caucasus, and Siberian Russia

Lesson 3: Life in the Regions

> ### ESSENTIAL QUESTION
> *How does geography influence the way people live?*

Terms to Know
oasis fertile area that rises in a desert wherever water is regularly available
homogenous of the same kind
yurt large round tent made of animal skins; used as a home in Central Asia

---

What Do You Know?

In the first column, answer the questions based on what you know before you study. After this lesson, complete the last column.

Now...		Later...
	What languages do people in Central Asia speak?	
	What religions do people in Central Asia practice?	
	What languages do people in the Caucasus speak?	
	What religions do people in Siberian Russia practice?	

networks

Central Asia, the Caucasus, and Siberian Russia

Lesson 3: Life in the Regions, *continued*

Marking the Text

1. Highlight the ethnic peoples who live in Siberia.

Defining

2. What is an *oasis*?

_____

_____

_____

_____

Making Connections

3. Why is there such a mix of ethnicities in Siberian Russia?

_____

_____

_____

_____

_____

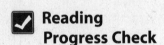

Reading Progress Check

4. Where does most of Central Asia's Russian population live?

_____

_____

_____

_____

People and Places

Guiding Question *Who are the peoples of Siberia, Central Asia, and the Caucasus?*

More than half of Siberians live in cities and towns, and most are near the Trans-Siberian Railroad. A few small cities and many towns also exist along rivers and Siberia's eastern coast. Siberia's population is mainly Russian. However, there are also Mongols and Turkic groups, as well as other indigenous peoples

In Kazakhstan most Russians and many Kazakhs live in Alma-Ata, the capital. About 40 percent of the country's people live in rural areas. Turkmenistan is evenly divided between city and rural dwellers. Most people live along the Amu Dar'ya and **oasis** areas to the south. An oasis is a green area by a water source in a dry region. Most Turkomans live in villages while most Russians live in the capital, Ashkhabad, and other cities.

In Uzbekistan, most people live in the eastern half of the country. Almost two-thirds live in rural areas. In Kyrgyzstan, less than 40 percent of the people live in cities and towns. Tajikistan is mostly rural, but the country's mountainous terrain causes its settled areas to be densely populated.

Central Asian countries have a variety of ethnicities, as shown in the table below.

Country	Ethnicity
Kazakhstan	66% ethnic Kazakhs; 25% Russians
Turkmenistan	75% ethnic Turkomans; 10% Russians
Uzbekistan	Uzbeks, Russians, and Kazakhs
Kyrgyzstan	66% Kirghiz; the rest are Uzbeks and Russians
Tajikistan	80% Tajiks; the rest are Uzbeks

The Caucasus region is evenly divided between urban and rural dwellers. In Armenia, about two-thirds of the people live in cities and towns with the Ararat Plain being the most heavily populated region. Most people in Armenia are ethnic Armenians.

Georgia's Black Sea coast is densely populated, and one of every four Georgians lives in the capital of Tbilisi. Georgia has many minorities, mainly Armenians and Azeris. There are also 50 smaller ethnic groups. Azerbaijan's most densely populated area is around its capital on the Caspian Sea. The most populated rural area is in the southeast. Most of its people are ethnic Azeris.

Central Asia, the Caucasus, and Siberian Russia

Lesson 3: Life in the Regions, *continued*

People and Cultures

Guiding Question *How are the cultures of Central Asia and the Caucasus alike and different?*

In Siberia, the main language is Russian. Russian is also the official language in Kazakhstan and Kyrgyzstan, along with the language of each country's ethnic majority. In the Caucasus, most Armenians and Georgians speak the language of their country's ethnic majority. Most Azerbaijanis speak their ethnic minority language as well.

The various religions practiced in countries of this region reflect its history of conquest by Arab, Chinese, and Russian empires.

Country	Religion
Siberia	Russian Orthodox Christianity; Buddhist and Muslim minorities
Kazakhstan	50% Christian; 50% Muslim
Armenia	mostly Christian
Azerbaijan	mostly Muslim
Georgia	mostly Christian; significant Muslim minority
remaining countries of Central Asia	75–90% Muslim; remainder Christian

Compared to the rest of the Caucasus, Armenians are **homogenous**, or of the same kind. Recently, Armenians and Azeris have been fighting over a region with an Armenian majority that is part of Azerbaijan.

In 1992 and 2008, ethnic minority groups in Georgia revolted, seking independence. Russia sent troops to help the rebels. The Russians withdrew, however, and the tensions continue today. In Central Asia, Soviet rule led to civil war.

Most urban Kazakhs live and dress like Europeans while rural Kazakhs wear traditional clothing. Many Kazakhs live in rug-filled **yurts**, a type of round tent with a domed roof. Most Kirghiz and many other rural Central Asians also live in yurts. The women stay home and care for their children while the men tend the family's crops and livestock.

Uzbek families are often large with close relationships. Most Uzbeks wear traditional clothing. In towns and cities they live in small houses with courtyards, in which they spend a great deal of time. Most Tajiks live in villages with flat-roofed houses along an irrigation canal or a river. A mud fence surrounds each house.

✏️ Marking the Text

5. Circle the countries in which Russian is the main or official language.

⚙️ Analyzing

6. How did Islam come to the countries of the Kazakhstan and Azerbaijan?

_____

_____

_____

🔤 Determining Word Meaning

7. What is a *yurt*? Which groups of people live in yurts?

_____

_____

_____

☑️ Reading Progress Check

8. What ethnic unrest exists in the Caucasus today?

_____

_____

_____

_____

_____

networks

Central Asia, the Caucasus, and Siberian Russia

Lesson 3: Life in the Regions, *continued*

Marking the Text

9. Read the paragraph on the Aral Sea. Circle the reason for the disappearance of the Aral Sea.

✓ Reading Progress Check

10. Why were the Caucasus and Central Asia slow to establish democratic governments when the Soviet Union collapsed?

_____

_____

_____

_____

Relationships and Challenges

Guiding Question *What challenges lie ahead for the regions?*

After the end of Soviet rule, most Caucasus and Central Asian countries struggled to establish stable, democratic governments. However, this was not always easy to do. Often, the same Soviet leaders or former Soviet officials kept power.

In addition, Central Asian countries struggle with Russian emigration. This has deprived some countries of workers in industries and businesses. Also, public health in Central Asia has suffered because of pollution and related health risks.

The Aral Sea is an example of an environmental disaster created by the Soviet Union. In the 1960s, the Soviet Union built huge farms in the desert and steppes. Then they dug long canals to irrigate them from the Aral Sea. As a result, the Aral Sea has lost 90 percent of its water. The seafloor is now dry land filled with salt and farm chemicals. Winds carry the salt and chemicals miles away and people breathe them in, increasing lung diseases and cancer rates.

Soviet rule also created border disputes between the Central Asian and the Caucasus countries. When the borders for these countries were created, the Soviets paid little attention to where the ethnic groups actually lived. In addition, several countries cannot agree on how to share the Caspian Sea. At stake is access to minerals, fishing rights, and transportation.

Writing

Check for Understanding

1. **Informative/Explanatory** Why does most of Siberia's population live along the Trans-Siberian railroad?

_____

_____

2. **Informative/Explanatory** What cultural characteristics do many of Central Asia's main ethnic groups share?

_____

_____

_____

Southwest Asia

Lesson 1: Physical Geography of Southwest Asia

ESSENTIAL QUESTION
How does geography influence the way people live?

Terms to Know
alluvial plain an area built up by rich fertile soil left by river floods
oasis an area in a desert where underground water allows plants to grow year-round
wadi a dry riverbed in a desert that fills with water when rains fall
semiarid somewhat dry

Where in the World: Southwest Asia

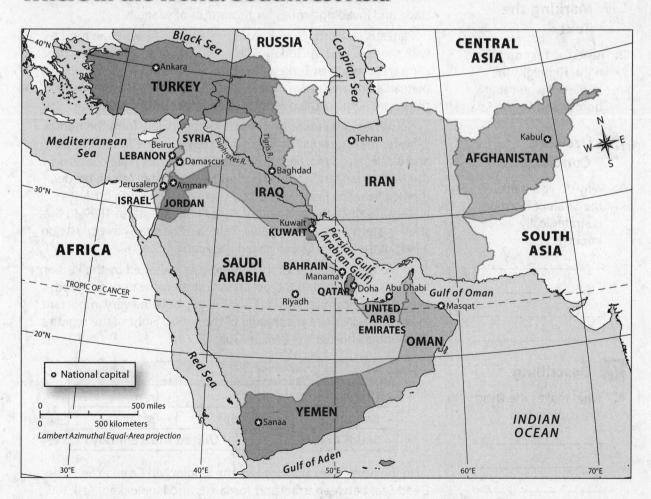

Lesson 1: Physical Geography of Southwest Asia, *continued*

? Explaining

1. Where is the Khyber Pass, and why is it important?

   _____
   _____
   _____
   _____
   _____

✎ Marking the Text

2. Read the text on the right. Highlight the bodies of water that border Southwest Asia.

⚙ Drawing Conclusions

3. Why do you think the Strait of Hormuz is strategically important?

   _____
   _____
   _____
   _____

? Describing

4. What makes the Dead Sea distinct?

   _____
   _____
   _____

Southwest Asia's Physical Features

Guiding Question *What are the main landforms and resources in Southwest Asia?*

Mountains and plateaus dominate the landscape of Southwest Asia. They have been created over the past 100 million years by collisions between four tectonic plates. This has also caused earthquakes in the region.

The region's highest mountains are the Hindu Kush range. It stretches across much of Afghanistan and along that country's border with Pakistan in South Asia. Mountain ranges form natural barriers to travel and trade. As a result, mountain passes are important in this area. One of the most famous is the Khyber Pass, between Afghanistan and Pakistan. It has served as a route for trade and invading armies for thousands of years.

Plateaus cover much of Iran. Mountains in western Iran merge with those of eastern Turkey. The Anatolian Plateau spreads across central and western Turkey. The Arabian Peninsula is a single, vast plateau. Long mountain ranges that lie parallel to the peninsula's coasts are actually the deeply eroded edges of the plateau.

Southwest Asia takes its physical shape mainly from the bodies of water that surround it. Turkey has coasts on the Mediterranean and Black Seas. Syria, Lebanon, Jordan, and Israel have coasts on the Mediterranean Sea. Jordan, Saudi Arabia, and Yemen border the Red Sea. The Red Sea has been one of the world's busiest waterways since the opening of Egypt's Suez Canal in 1869. In the southeastern Arabian Peninsula, Yemen and Oman have coasts on a part of the Indian Ocean called the Arabian Sea.

In the northeast, the Arabian Peninsula is shaped by the Persian Gulf. It is connected to the ocean by a strategic waterway called the Strait of Hormuz. The Persian Gulf has become very important in world affairs since the middle of the 1900s. Eight of the region's 15 countries border the Persian Gulf.

Persian Gulf Countries		
• Oman	• Iraq	• Iran
• Bahrain	• Kuwait	• Qatar
• Saudi Arabia	• United Arab Emirates	

In the north, Iran also borders the landlocked Caspian Sea. The Dead Sea, between Israel and Jordan, is also landlocked. At 1,300 feet (395 m) below sea level, it is the world's lowest body of water. Its shore represents the world's lowest land elevation.

Southwest Asia

Lesson 1: Physical Geography of Southwest Asia, *continued*

Southwest Asia's two longest and most important rivers are the Tigris and the Euphrates. They are often considered part of the same river system. They begin in the mountains of eastern Turkey. In their lower courses, they flow parallel to each other through an **alluvial plain**. This is a plain created by sediment deposited during floods. The plain covers most of Iraq as well as eastern Syria and southeastern Turkey. This area has been known since ancient times as Mesopotamia, which is Greek for "land between the rivers."

Deserts spread across much of Southwest Asia. The Arabian Desert covers nearly the entire Arabian Peninsula. It is the largest desert in the region, and includes many landscapes, including sand seas, which are unbroken expanses of sand. The largest sand sea in the world is the Rub' al-Khali, or Empty Quarter, in the southern part of the peninsula. The climate is so dry that there are no permanent human settlements. In some areas, nomadic people known as the Bedouin keep herds of camels, horses, and sheep. There are **oases** in the Arabian Desert. These are areas where underground water allows plants to grow.

Southwest Asia's Climates

Guiding Question *What are some ways that mountains, seas, or other physical features affect climate in Southwest Asia?*

Most of Southwest Asia has an arid, or very dry, climate. Deserts cover nearly the entire Arabian Peninsula as well as large parts of Iran. Temperatures can rise to 129°F (54°C). When rains come, torrents of water race through **wadis**, or streambeds that are dry much of the year. Seeds sprout within hours, turning the plains green. Areas along the edges of the dry zones are **semiarid**, or somewhat dry. These are found in highlands and mountain ranges.

Along the region's Mediterranean and Aegean coasts and across much of western Turkey, the climate is Mediterranean. There are mild temperatures and moderate rainfall in the winter. Summers are warm and dry. The mountains of eastern Turkey, western Iran, and central Afghanistan have continental climates. Temperatures vary greatly between summer and winter. The mountains of the Hindu Kush have a highland climate. Glaciers are found among their high peaks.

Natural Resources

Guiding Question *How do natural resources influence the lives of people in Southwest Asia?*

Scarcity of water has shaped Southwest Asia's human history and settlement patterns. The region's most important resources are oil and natural gas.

? Identifying

5. What is the Rub' al-Khali?

_____

_____

_____

✓ Reading Progress Check

6. How has tectonic activity helped shape landforms in Southwest Asia?

_____

_____

_____

A♭c Defining

7. Describe the difference between an *oasis* and a *wadi*.

_____

_____

_____

✓ Reading Progress Check

8. In what parts of Southwest Asia can farmers probably grow crops without irrigation?

_____

_____

_____

Southwest Asia

Lesson 1: Physical Geography of Southwest Asia, *continued*

 Marking the Text

9. Read the text on the right. Highlight recent discoveries in Afghanistan that may help it rise out of poverty.

✓ **Reading Progress Check**

10. Five countries that border the Persian Gulf hold more than half the oil that has been discovered in the world. Name three of the countries.

_____

_____

_____

Natural Gas	• Gaseous form of petroleum
Crude Oil	• Liquid form of petroleum • Gasoline; diesel, heating, and industrial oil; plastics, cloth fibers

The world's largest known deposits of petroleum are in Southwest Asia. Most of them are concentrated around and under the Persian Gulf. Five countries that border the gulf—Saudi Arabia, Iran, Iraq, Kuwait, and United Arab Emirates—hold more than half the world's known oil. These countries export most of their oil to industrialized countries. Petroleum revenues have brought tremendous wealth to a few people in the exporting countries. Only in a few areas, how ever, has the wealth been used to improve the lives of the people or to bring about modernization.

The region boasts a great variety of mineral resources. Turkey and Iran have significant deposits of coal. Phosphates, which are used in fertilizers, are mined in Iraq, Israel, and Syria. Between 2006 and 2010, American geologists in Afghanistan discovered enormous deposits of many minerals. Among them were iron, copper, gold, cobalt, lithium, and rare earth elements used to make electronic devices.

Writing

Check for Understanding

1. **Informative/Explanatory** What are the major physical features of Southwest Asia?

_____

_____

2. **Informative/Explanatory** How has petroleum affected the countries that export it?

_____

_____

_____

networks

Southwest Asia

Lesson 2: History of Southwest Asia

ESSENTIAL QUESTION

Why do civilizations rise and fall?

Terms to Know

polytheism the belief in more than one god
millenium a period of a thousand years
monotheism the belief in one god

When did it happen?

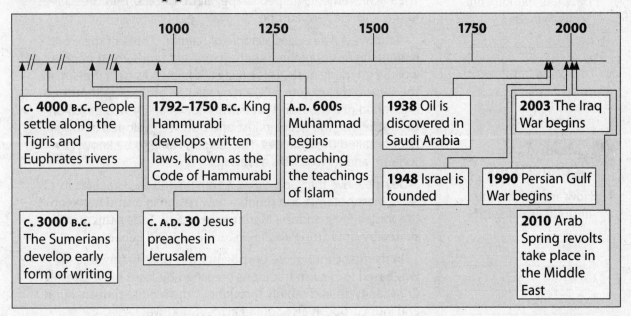

| 1000 | 1250 | 1500 | 1750 | 2000 |

c. 4000 B.C. People settle along the Tigris and Euphrates rivers

1792–1750 B.C. King Hammurabi develops written laws, known as the Code of Hammurabi

A.D. 600s Muhammad begins preaching the teachings of Islam

1938 Oil is discovered in Saudi Arabia

2003 The Iraq War begins

c. 3000 B.C. The Sumerians develop early form of writing

c. A.D. 30 Jesus preaches in Jerusalem

1948 Israel is founded

1990 Persian Gulf War begins

2010 Arab Spring revolts take place in the Middle East

Early Southwest Asia

Guiding Question *What are some of the most important advancements that occurred in Southwest Asia in ancient times?*

Throughout most of history, people have lived as hunter-gatherers. In small groups, they hunted wild animals and searched for wild fruits, nuts, and vegetables. About 10,000 years ago, people began practicing agriculture—raising animals and growing crops. One of the first places this agricultural revolution took place was in Mesopotamia. Mesopotamia is the fertile plain between the Tigris and the Euphrates Rivers in what is now Iraq.

? Describing

1. What change began to take place in Mesopotamia around 10,000 years ago?

_____

_____

Southwest Asia

Lesson 2: History of Southwest Asia, *continued*

Marking the Text

2. Read the text on the right. Highlight the sentence that tells how the world's first civilizations developed.

Defining

3. What is the difference between *monotheism* and *polytheism*?

_____

_____

_____

_____

Making Connections

4. How are Christianity and Judaism linked?

_____

_____

_____

_____

Making Connections

5. What do Christianity, Islam and Judaism have in common?

_____

_____

_____

_____

With agriculture came a more settled life. Villages began to appear. Because food was plentiful, some villagers were freed up from farming and could undertake other work, such as making tools, weaving baskets, or keeping records. Over time some villages grew into large, powerful cities with governments and military forces. These cities were the world's first civilizations.

Over thousands of years, Mesopotamian societies such as the Sumerians and the Babylonians invented sophisticated irrigation and farming methods. They built impressive works of architecture, including huge, temple towers. They made advances in mathematics, astronomy, government, and law. They also created a writing system and produced great works of literature. Their achievements helped shape later civilizations in Greece, Rome, and Western Europe.

Southwest Asia is also a cradle of religion. Three of the world's major religions originated there. In ancient times, most people worshiped many gods. This practice is known as **polytheism**. In the second millennium B.C., a new religion arose. A **millennium** is a period of a thousand years. This new religion was based on **monotheism**, the belief in just one God. It developed among a people called the Israelites. The religion came to be known as Judaism and its followers as Jews.

Jews believe that God called Abraham and the Israelites to leave Mesopotamia and found a new nation in a land between the Jordan River and the Mediterranean Sea. Today, this area is shared by Israel, the Palestinian territories, and Lebanon.

In the first century A.D., Christianity was born in this area. It was based on Jewish traditions preached by Jesus of Nazareth. Christianity spread rapidly throughout the Mediterranean world and into Europe. It also spread across Southwest Asia.

Then, in the A.D. 600s, the prophet Muhammad founded Islam in Makka (Mecca), a desert city in what is now Saudi Arabia. Islam is the religion of Muslims. Many of the teachings of Islam are similar to those of Judaism and Christianity. For example, all three are monotheistic and consider Abraham to be God's first messenger who first taught this belief.

Three World Religions from Southwest Asia			
Religion	**Judaism**	**Christianity**	**Islam**
Founder	Abraham	Jesus of Nazareth	Muhammad
Followers	Jews	Christians	Muslims

Southwest Asia

Lesson 2: History of Southwest Asia, *continued*

Muhammad's religion was relatively simple and direct. It focused on the need to obey the will of Allah, the Arab word for God. It required followers to perform five duties, known as the Pillars of Islam. These are: (1) promising faith to God and accepting Muhammad as God's prophet; (2) praying five times a day; (3) fasting during the holy month of Ramadan; (4) helping the poor; and (5) making a journey to the holy city of Makkah

In its first years, Islam attracted few converts. By the time of Muhammad's death in A.D. 632, it had spread across the Arabian Peninsula. Arab armies began spreading the religion through conquests. It was also spread by scholars, religious pilgrims, and Arab traders. By A.D. 800, Islam had spread across nearly all of Southwest Asia including Persia (present-day Iran) and part of Turkey. It also extended into northernmost Africa and most of Spain and Portugal. It later expanded to northern and eastern Africa, Central Asia, and South and Southeast Asia.

Many different influences contributed to a flowering of Islamic culture that lasted for centuries and continues to have an impact. During this period, great works of architecture were built, and centers of learning arose. Arab scholars made advances in math and science. This golden age was to have a lasting impact.

During the 1100s and 1200s, crusaders from Western Europe set up Christian states along Southwest Asia's Mediterranean coast. The Muslims gradually regained control of these territories. In other areas, however, Muslim political and military power weakened. In the 1200s, Persia and Mesopotamia became part of a vast Mongol empire that stretched across much of Eurasia. The Ottomans, a group of Muslim tribes, began to build an empire on the Anatolian Peninsula in the early 1200s. By the mid-1300s, the Ottoman Empire included much of Southwest Asia and parts of southeastern Europe and northern Africa. It became one of the world's most powerful states.

Modern Southwest Asia

Guiding Question *What present-day issues facing Southwest Asia have their roots in ancient times?*

The Ottoman Empire reached the peak of its power in the 1500s, then began to decline. In the 1800s and early 1900s, it lost African and European territories through wars, treaties, and revolutions. After fighting alongside the losing Central Powers in World War I, the empire was finally dissolved. After World War I, the modern country of Turkey was founded on the Anatolian Peninsula.

? Identifying

6. What are the Pillars of Islam?

_____

_____

_____

_____

_____

✎ Marking the Text

7. Read the text on the left. Underline the sentences that describe the flowering of Muslim culture.

? Locating

8. When and where did the Ottoman Empire begin?

_____

_____

_____

_____

☑ Reading Progress Check

9. What are some ways in which Islam was spread?

_____

_____

_____

Southwest Asia

Lesson 2: History of Southwest Asia, *continued*

Marking the Text

10. Read the text on the right. Highlight the names of European countries that received territory in Southwest Asia under the mandate system following World War I.

Analyzing

11. How did the political boundaries of the former Ottoman Empire territories cause problems later?

_____

_____

_____

Defining

12. What was the Holocaust?

_____

_____

_____

Analyzing

13. What are the roots of conflict between Arabs and Jews in the region?

_____

_____

_____

_____

In the peace settlement that ended the war, Britain and France gained control of the Ottoman Empire's former territories under a mandate system. The people of these territories were to be prepared for independence. The new political boundaries did not respect ethnic, religious, political, or historical divisions. These boundaries would become very important when the territories became independent countries. Resentment toward the European colonial powers grew into strong nationalist movements among the peoples of the region. Between 1930 and 1971, one country after another won independence.

One of the British mandates was the territory called Palestine. It roughly corresponded to Israel, the area inhabited by the Jewish people in ancient times. Most of the people living in Palestine were Muslim Arabs. At the same time, growing numbers of Jewish immigrants had been arriving from Europe and other parts of the world. As the Jewish population grew, tensions increased between them and the Palestinian Arabs.

Jewish nationalists wanted to re-establish their historic homeland in Palestine. This movement gained support as a result of the Holocaust. The Holocaust was the systematic murder of 6 million European Jews by Nazi Germany during World War II. Hundreds of thousands of Jews who survived became refugees, searching for a place to live.

In 1947 the United Nations voted to divide the territory into two states, one Arab and one Jewish. The Arabs did not want to give up land, and rejected the proposal. On its independence day in 1948, the new Jewish state of Israel was invaded by armies from neighboring Arab countries. Hundreds of thousands of Palestinian Arabs became refugees. That war ended with a truce in 1948. However, other major Arab-Israeli wars were fought in the next three decades. During a brief 1967 war, Israel captured territory in neighboring countries.

Areas Captured by Israel in 1967 War
West Bank • East Jerusalem • Sinai Peninsula
Gaza Strip • Golan Heights

Opposition by Palestinian Arabs and neighboring countries led to further conflict. Israel withdrew from the Sinai Peninsula in 1982 and from the Gaza Strip in 2005. It continues to control the West Bank and East Jerusalem. There have been many unsuccessful attempts to find a peaceful solution to the Arab-Israeli conflict.

Southwest Asia

Lesson 2: History of Southwest Asia, *continued*

Since World War II, ethnic, religious, and political differences have caused many conflicts. Islamists consider Islam to be a political system as well as a religion. Islamist movements and the desire to control large oil fields have contributed to many of the conflicts. Lebanon, Afghanistan, and Yemen have been torn apart by civil wars. The Kurds are a people in eastern Turkey, northern Iraq, and western Iran. They have fought to gain their own country. A revolution in Iran in the late 1970s overthrew the monarchy and created an Islamic republic. Iraq invaded Iran in 1980, resulting in an eight-year war. A decade later, Iraq annexed the oil-rich neighboring country of Kuwait. A coalition led by the United States liberated Kuwait in the Persian Gulf War.

In response to terrorist attacks on U.S. soil that took place on September 11, 2011, the United States invaded Afghanistan. They removed the country's Islamist ruling group, the Taliban, from power. The Second Persian Gulf War began two years later, when U.S. and British forces invaded Iraq and overthrew Iraq's leader, Saddam Hussein. He had been accused of supporting terrorism and possessing weapons of mass destruction, although this was later proved to be untrue.

In 2010 and 2011, a popular uprising in Tunisia inspired democratic movements in Yemen, Bahrain, and Syria. However, militant Islamic political movements limit growth of democracy and civil rights in the region.

⚙ Describing

14. Who are the Kurds, and why are they involved in conflict?

_____

_____

_____

☑ Reading Progress Check

15. What event in Europe helped spur the creation of a Jewish state in Southwest Asia?

_____

_____

_____

Writing

Check for Understanding

1. **Informative/Explanatory** List three advancements that occurred in Southwest Asia in ancient times.

_____

_____

2. **Informative/Explanatory** List three major conflicts in Southwest Asia since World War II.

_____

_____

_____

netw✺rks

Southwest Asia

Lesson 3: Life in Southwest Asia

ESSENTIAL QUESTION
How does religion shape society?

Terms to Know
hydropolitics the politics related to water access and usage
fossil water water that fell as rain thousands of years ago and is now trapped deep below ground

What Do You Know?

In the first column, answer the questions based on what you know before you study. After this lesson, complete the last column.

Now...		Later...
	What major ethnic groups live in Southwest Asia?	
	How has petroleum changed life in Southwest Asia?	
	What is the Arab Spring?	

? Identifying

1. What are Southwest Asia's two most populous countries, and approximately how many people live in each country?

_____

_____

_____

_____

_____

People and Places

Guiding Question *In what parts of Southwest Asia do most people live?*

Southwest Asia is about three-fourths the size of the United States but is home to around 330 million people. Iran and Turkey, the most populous countries, each have about 80 million people. Some oil-rich countries around the Persian Gulf are experiencing population booms as their fast-growing economies are attracting foreign workers. Many countries are highly urbanized. In others, the majority of people live in rural areas. People below the age of 15 make up a large part of the population here.

   Population is not evenly distributed. The region's northern and western parts and southern tip have the highest densities. Many of these areas have relatively higher rainfall. Areas with dry or somewhat dry climates are more sparsely populated. One exception is Mesopotamia in Iraq, where the rivers provide abundant water for irrigating crops.

Southwest Asia

Lesson 3: Life in Southwest Asia, *continued*

Southwest Asia has thriving metropolises that are home to millions. It also has ancient rural villages that seem untouched by the passing of time. In some desert areas, nomadic people known as Bedouins live in tents and herd camels, goats, and cattle.

People and Cultures

Guiding Question *What cultural differences are found across Southwest Asia?*

Arabs are the largest ethnic group in Southwest Asia. In Saudi Arabia, Syria, and Jordan, more than 90 percent of the people are Arab. The two most populous countries have only small Arab populations. In Turkey, Turks form the majority. In Iran, which was formerly known as Persia, most people are Persian. Jews make up three-fourths of the population in Israel. The Kurds have no country of their own, but make up significant minorities in Turkey, Iran, and Iraq. The region where they live is traditionally known as Kurdistan.

Arabic is the most widespread language in Southwest Asia. Other important tongues include Turkish and Farsi, the language of Persians. Hebrew is the official language of Israel. Kurds speak Kurdish. Some of the region's countries have complex ethnic and linguistic makeups. Many people identify with their ethnic group more strongly than with the country in which they live. Afghanistan for example has many ethnic groups.

Afghanistan	
Ethnic Groups	**Languages**
• Pashtuns	*Official*:
• Tajiks	• Afghan Persian
• Hazaras	• Pashto
• Uzbeks	*Spoken*:
• Aimaks	• more than 30 other languages
• Turkmen	
• Balochs	

People in Afghanistan identify themselves as Pashtun or Hazari rather than Afghani. Even in countries that are mostly Arab, people identify with tribes that are based on family relationships. Tribal identity is often stronger than national identity.

? Describing

2. Who are the Bedouin?

_____

_____

_____

✓ Reading Progress Check

3. Why do some countries around the Persian Gulf have rapidly growing populations?

_____

_____

_____

_____

  Marking the Text

4. Read the text on the left. Underline sentences that refer to the people and culture of Israel.

⚙ Determining Central Ideas

5. Why has it been difficult to create a sense of national identity in some Southwest Asian nations?

_____

_____

Reading Essentials and Study Guide **217**

Southwest Asia

Lesson 3: Life in Southwest Asia, *continued*

? **Identifying**

6. What are the two main branches of Islam? To which branch do most Southwest Asians belong?

_____

_____

 Marking the Text

7. Read the text on the right. Underline the sentence that tells what areas in life are governed by Islam.

✓ **Reading Progress Check**

8. What is the major ethnic group in Iran? What language does that group speak?

_____

_____

⚙ **Drawing Conclusions**

9. How might Persian Gulf countries be affected if countries began turning to alternate energy sources?

_____

_____

_____

Islam remains the region's dominant religion. It unites people of different ethnicities and languages. It has two main branches, Sunni and Shia. Most of Southwest Asia's Muslims are Sunnis. In Iran, however, Shias outnumber Sunnis nine to one. Judaism is practiced by three-fourths of the people in Israel. Christians make up about 40 percent of the population in Lebanon and 10 percent in Syria.

Religion and art have always been closely tied in the region. Some of its most magnificent works of architecture are religious structures. Sacred texts such as the Hebrew and Christian Bible and the Muslim Quran are also works of literature. There are also other rich artistic traditions, including calligraphy, mosaics, weaving, storytelling, and poetry. Handwoven Persian carpets have been famous for centuries. So has the collection of folktales known as *The Thousand and One Nights.*

Daily life varies greatly across the region. Throughout history, most people practiced traditional livelihoods such as farming, raising livestock, or fishing. In recent times, people have gone to work in petroleum production, food processing, auto manufacturing, textiles, and construction.

Religion plays a major role in the daily lives of many people in Southwest Asia. Islam is a way of life. There are rules governing diet, hygiene, relationships, business, law, and more. To Muslims, families are the foundation of a healthy society. Maintaining family ties is an important duty.

Ramadan, the ninth month of the Muslim calendar, is a holy month of fasting. Muslims refrain from eating and drinking between dawn and dusk. After ending their fast with prayer each evening, people enjoy festive meals. The end of Ramadan is marked by a three-day festival called *Eid al-Fitr.*

Issues

Guiding Question *How have oil wealth and availability of natural resources created challenges for countries of Southwest Asia?*

The discovery in the mid-1900s of vast petroleum deposits in Southwest Asia had a strong impact on the region. Exports of petroleum products have brought great wealth to countries around the Persian Gulf. With this wealth came development in some countries. Petroleum has also brought new challenges. Some Muslims believe that the increased exposure to Western ways is corrupting the region's people. There is a growing gap between rich and poor countries. The struggle to control oil has led to tension and wars. It has also brought foreign intervention.

Southwest Asia

Lesson 3: Life in Southwest Asia, *continued*

The years 2010 and 2011 marked the beginning of the Arab Spring. This was a wave of pro-democracy protests in North Africa and Southwest Asia. By the end of the year, leaders in Tunisia, Egypt, Libya, and Yemen were overthrown. Syria fell into upheaval.

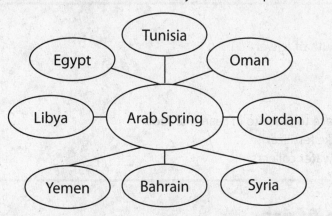

The scarcity of freshwater has plagued the region throughout history. Dramatic population growth has greatly increased demand, making the situation more dire. This has increased the importance of **hydropolitics**, or politics related to water usage and access. Salt can be removed from the seawater, but this process is expensive and therefore impractical.

With no rivers that flow year-round, Saudi Arabia has tapped into **fossil water**. This is water that fell thousands of years ago as rain and is now trapped deep underground. This is not a renewable source, however. The region's greatest source of freshwater is the Tigris-Euphrates river system. However, a dam-building project in Turkey threatens to reduce river flow to Syria and Iraq.

⚙ Analyzing

10. How might dams built on the Tigris and Euphrates rivers in Turkey affect agriculture in Syria and Iraq?

_____

_____

✔ Reading Progress Check

11. What was the Arab Spring? What countries in Southwest Asia were involved?

_____

_____

_____

_____

_____

Writing

Check for Understanding

1. Informative/Explanatory How has religion shaped society in Southwest Asia?

_____

_____

2. Informative/Explanatory How have hydropolitics affected Southwest Asia?

_____

_____

North Africa

Lesson 1: Physical Geography of North Africa

ESSENTIAL QUESTION

How does geography influence the way people live?

Terms to Know

delta area formed by soil deposits at the mouth of a river
silt fine, rich soil good for farming
wadi dry desert riverbed
erg vast stretches of sand
nomad person who lives by moving from place to place to in search of food
phosphate chemical compound used to make fertilizer
aquifer underground layer of rock in which water collects

Where in the World: North Africa

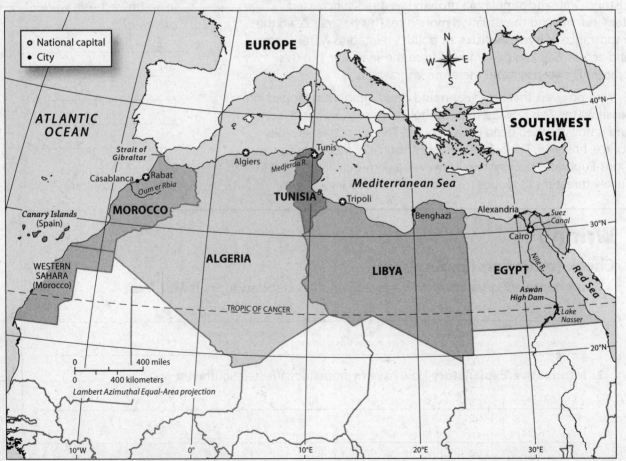

North Africa

Lesson 1: Physical Geography of North Africa, *continued*

Landforms and Waterways

Guiding Question *How have physical features shaped life in the region?*

North Africa includes five countries. Egypt is the easternmost country in the region. Some of Egypt is in Southwest Asia. Libya is to Egypt's west. Tunisia and Algeria are west of Libya, and Morocco is farthest to the west.

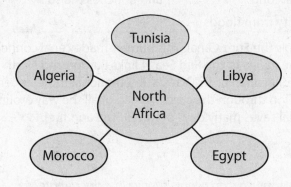

Low, narrow plains sit on the edges of the region's Atlantic and Mediterranean coasts. In the west, the Atlas Mountains rise behind these coastal plains. The mountains extend across Morocco and Algeria into Tunisia. South of the Atlas Mountains is a low plateau that stretches across most of North Africa.

In Egypt, the southern part of the Nile River cuts through a highland area. There it forms a deep gorge, or valley. Southeastern Egypt has low mountains on the shores of the Red Sea. The southern part of Egypt's Sinai Peninsula is also mountainous. Northwestern Egypt has a large area of lowland that contains marshes and lakes.

After the Mediterranean Sea, the most important body of water in the region is the Nile River. At 4,160 miles (6,695 km), it is the longest river in the world. It begins far south of Egypt at Lake Victoria in East Africa. Then it flows northward and is joined by several tributaries, including the Blue Nile.

The Nile has a large **delta** at its mouth on the Mediterranean Sea. A delta is an area formed by soil deposits that build up as a river's waters slow down. Deltas often form where a river enters a larger body of water.

The Nile brings life to dry Egypt. In ancient times, it flooded each year. These floods left **silt** along the banks of the river and in the delta. Silt is a fine, rich soil that is excellent for farming. Because farmers could grow food there, they were able to support a great civilization. Ancient Egypt was called "the gift of the Nile."

✏ Marking the Text

1. Highlight all the areas in North Africa that have mountains.

❓ Identifying

2. What bodies of water border North Africa?

   _____

   _____

❓ Describing

3. Describe the course of the Nile through North Africa.

   _____

   _____

   _____

   _____

🅰🅱🅲 Defining

4. What is a *delta*?

   _____

   _____

   _____

❓ Explaining

5. How do farmers benefit from floods along the Nile River?

   _____

   _____

   _____

Lesson 1: Physical Geography of North Africa, *continued*

❓ Explaining

6. Why is the Suez Canal important?

_____

_____

_____

✅ Reading Progress Check

7. Why was ancient Egypt called "the gift of the Nile"?

_____

_____

_____

✏️ Marking the Text

8. Underline the text that explains the rain shadow effect.

🔤 Defining

9. What is a *wadi*?

_____

_____

✅ Reading Progress Check

10. Where do you think most people in North Africa live? Why?

_____

_____

_____

Today, several dams control floods on the Nile. They hold back the high volume of water during rainy season. Then they release water during the rest of the year. There are advantages and disadvantages to the dams.

Advantages	Disadvantages
• Crops can be grown all-year long • Security from floods	• Silt no longer settles and enriches the soil

Egypt controls the Suez Canal. This human-made canal connects the Mediterranean Sea to the Red Sea. It links Europe and North Africa to the Indian and Pacific Oceans. International trade depends on ships travelling through the canal instead of all the way around Africa. The canal saves many days of travel time and fuel.

Climate

Guiding Question *How do people survive in a dry climate?*

Inland North Africa is dry. One reason is the rain shadow effect. This is when moist air blows south and rises up the Atlas Mountains, cooling the air and releasing rain. By the time the air passes over the mountains, it is dry. Also high-pressure air systems fall over areas to the south, sending hot, dry air north. When it does rain in the desert, southern winds soon follow. These winds dry the land and leave behind **wadis**, or dry streambeds.

Much of North Africa is covered by the Sahara. The Sahara's vast stretches of sand are called **ergs**. Strong winds blow the sand, creating dust storms and sand dunes. Ergs cover only about a quarter of the Sahara. There are also rocky plateaus calls hamadas, rocks eroded by wind, and oases.

Trade caravans and **nomads** rely on oases. Nomads are people who move about from place to place in search of food. They use plants in oases to graze herds of sheep or other animals. Some people live on oases and grow crops.

In the north, a band of steppe area encircles the desert from coast to coast. In the steppe, temperatures are high and rainfall is slightly greater than in the desert.

A Mediterranean climate dominates the western coast. This gives the region warm, dry summers and mild, rainy winters. More rain falls along the coast and on the mountain slopes. Highland climates are found in the mountains. Morocco's Atlas Mountains can even get snow in the winter.

North Africa

Lesson 1: Physical Geography of North Africa, *continued*

Resources

Guiding Question *What resources does North Africa have?*

Libya is the most oil-rich country in North Africa. It has the ninth-largest amount of oil reserves in the world. It also has smaller amounts of natural gas. Algeria has large reserves of both natural gas and oil. Like Algeria, Egypt has larger reserves of natural gas than oil.

Tunisia's main resources are iron ore and **phosphates**. Phosphates are chemical compounds often used in fertilizers. Morocco also has phosphates, as well as rich fishing grounds off its coast.

All five North African countries struggle to get enough water. Limited rainfall and high temperatures leave little fresh water on the surface. Rains can be heavy when they come, but the sandy soil soon absorbs the water. Dry winds evaporate the rest. Only the Nile is a reliable source of water for farming. Ninety-five percent of Egyptians live within 12 miles of the Nile or its delta. They could not survive without its water.

Outside the Nile valley, most of the water comes from oases and **aquifers**. Aquifers are underground layers of rock in which water collects. People use wells to tap into this water. Libya, for instance, relies on aquifers for almost all of its water needs.

Demand for the water in an aquifer shared by Algeria, Libya, and Tunisia has significantly increased in recent years. In North Africa, aquifers take a long time to fill up again. If people keep taking water out at a high rate, the aquifers could empty. Then the water problem will become much worse.

 Marking the Text

11. Circle the word that describes a chemical used in fertilizer.

? Explaining

12. How do people access the water in *aquifers*?
_____
_____

☑ Reading Progress Check

13. Why would aquifers take a long time to fill up in North Africa?
_____
_____
_____
_____

Writing

Check for Understanding

1. **Informative/Explanatory** Which countries in the region are not likely to need to import energy resources? Why?

_____

_____

2. **Informative/Explanatory** What sources of water do North Africans rely on and how are those sources threatened?

_____

_____

North Africa

Lesson 2: The History of North Africa

ESSENTIAL QUESTION

Why was ancient Egypt important?

Terms to Know

pharaoh king in ancient Egypt

myrrh substance from certain plants that gives off a pleasing scent

hieroglyphics system of writing that used pictures to represent sounds or words

convert change

monotheism belief in one God

caliph political and religious ruler of the Muslim empire

regime style of government

fundamentalist person who believes that government should strictly follow religious law

civil war fight between opposing groups for control of a country's government

When did it happen?

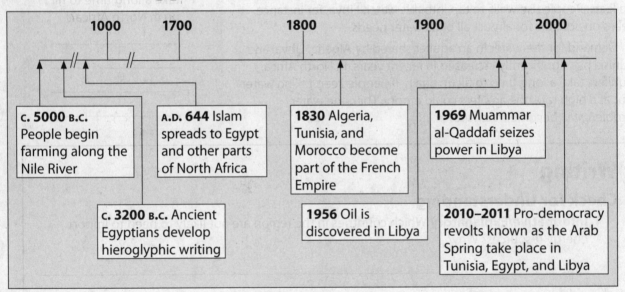

| 1000 | 1700 | 1800 | 1900 | 2000 |

c. 5000 B.C. People begin farming along the Nile River

c. 3200 B.C. Ancient Egyptians develop hieroglyphic writing

A.D. 644 Islam spreads to Egypt and other parts of North Africa

1830 Algeria, Tunisia, and Morocco become part of the French Empire

1956 Oil is discovered in Libya

1969 Muammar al-Qaddafi seizes power in Libya

2010–2011 Pro-democracy revolts known as the Arab Spring take place in Tunisia, Egypt, and Libya

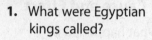

North Africa

Lesson 2: The History of North Africa, *continued*

Ancient Egypt

Guiding Question *Why was ancient Egypt important?*

As far back as 8,000 years ago, people settled along the banks of the Nile River. The rich flood waters helped farmers grow enough food to support a growing population. Over time, people began to do things besides farm. They made pottery, crafted jewelry, and became fighters. A few became kings.

About 5,000 years ago, two kingdoms along the Nile were united. For most of the next 3,000 years, kings called **pharaohs** ruled the land. Their people farmed the land and paid a share of their crops to the government. They also built pyramids and monuments for the pharaohs and fought in their armies.

For centuries Egypt traded with other lands. In the south, they traded grain for luxury goods like gold, ivory, and incense. In the east, they traded for wood. Around 1500 B.C., Egypt took control of lands to the south that held gold. They also seized areas along the Red Sea that had **myrrh**. This is a substance from certain plants that gives off a pleasing scent. Egypt also conquered the eastern shores of the Mediterranean because of the timber there.

The Egyptians believed in many gods. The sun god was an important god. His journey across the sky brought warmth that would help grow crops. They believed the pharaoh was the head of Egyptian society and the son of the sun god.

Egyptians also believed in life after death. Because of this, pharaohs had vast tombs filled with riches, food, and other goods, built for themselves. When a pharaoh died, his body was preserved as a mummy and placed in a tomb.

Historians know much about ancient Egypt because the Egyptians had a system of writing. This system was called **hieroglyphics**. It used pictures to represent sounds or words. Later, Egypt built one of the world's earliest libraries. It stored many important works of ancient literature.

Egyptians made many advances in technology, as well. Their knowledge spread to other lands through trade and conquest. Some of these advances are shown in the chart below.

Mathematics	Measured farm fields; figured out taxes
Study of stars and planets	Advances in astronomy
Engineering	Built great pyramids and temples

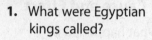

Defining

1. What were Egyptian kings called?

_____

Marking the Text

2. Highlight the passage that describes the lands conquered by the Egyptians. Where did they find myrrh? Where did they find timber?

_____

_____

Explaining

3. What happened to a pharaoh's body after he died?

_____

_____

_____

Reading Progress Check

4. Why is it important to know about ancient Egypt?

_____

_____

_____

_____

North Africa

Lesson 2: The History of North Africa, *continued*

Determining Cause and Effect

5. What happened in North Africa after the wars between Carthage and Rome in the 200s and 100s B.C.?

_____

_____

_____

Marking the Text

6. Circle the word that describes the belief in one god.

☑ **Reading Progress Check**

7. Did the Roman or Islamic empire have more impact on North Africa? Why do you think so?

_____

_____

_____

_____

Activating Prior Knowledge

8. Why was the Suez Canal important?

_____

_____

_____

The Middle Ages

Guiding Question *How was North Africa connected to other areas?*

Western North Africa was first visited by other Mediterranean peoples in the 600s B.C. Traders from what is now Lebanon built settlements there, including the city of Carthage. Within 200 years, Carthage controlled North Africa from modern Tunisia to Morocco. It also ruled parts of Spain and Italy.

After fighting wars with Rome in the 200s and 100s B.C., Carthage was destroyed. Rome came to control western North Africa and Egypt. In time, many North Africans **converted**, or changed, to Christianity, which was the religion of the Roman Empire. Otherwise the Roman Empire had little effect on North Africans.

In the A.D. 400s, the Roman Empire fell, and several kingdoms formed in North Africa. In the A.D. 600s, the religion of Islam emerged. Islam is a monotheistic religion. **Monotheism** means belief in one god. Followers of Islam, called Muslims, began to conquer other lands. By A.D. 705, they ruled all of North Africa.

The **caliph** was the political and religious authority of the Muslim empire. However, caliphs had trouble keeping control of North Africa. By the A.D. 800s, separate Berber kingdoms arose. In the A.D. 1000s, an Islamic group known as the Fatamids took power in Egypt. By the A.D. 1000s, most Berbers and Egyptians had converted to Islam and adopted the Arabic language.

The Modern Era

Guiding Question *What leads people to revolt against a government?*

In the 1500s, North Africa fell under the rule of foreign armies, including Portugal, Spain, and the Ottoman Empire. As Ottoman power weakened in the 1800s, Europeans moved into the area. At first, Egypt remained free. Once Egypt built the Suez Canal in 1860, however, Britain sent troops to gain control of the country. The table below shows when foreign countries took control of North African countries.

North African Country	European Power	Time Period
Algeria, Tunisia	France	Late 1800s
Morocco	France, Spain	Early 1900s
Libya	Italy	Early 1900s
Egypt	Britain	1882

networks

North Africa

Lesson 2: The History of North Africa, *continued*

Independence movements arose across North Africa, starting in the early 1900s. In 1952, Egypt became an independent republic. In 1962, Algeria ousted the French. In 1969, Libyan military leaders took control of the country. They were led by Muammar al-Qaddafi, who remained in power for 40 years. In 1959, Tunisia became a republic. In 1956, Morocco became a monarchy.

Independence did not end the region's problems, though. Algeria has been plagued by unrest among Islamic political groups. Tunisia's government has neglected the rights of its people. In Libya Qaddafi created a harsh **regime**, or style of government.

Other problems in the region included high population growth, corrupt governments, and the rise of Muslim **fundamentalists**. These are Muslims who want the government to follow the strict laws of Islam. They also reject Western influences on Muslim society.

In late 2010, a series of revolts called the Arab Spring took place in the region. In Tunisia, protests caused the president to step down. Emboldened by events in Tunisia, Egyptians protested until Hosni Mubarak, their longtime president, also stepped down in February 2011. In Morocco, the king agreed to several reforms. In Libya, a **civil war**, or a fight for control of the government, broke out. In October 2011, the rebels took control of the country and killed Qaddafi.

✎ Marking the Text

9. Underline problems North African countries dealt with after independence.

☑ Reading Progress Check

10. How did the reaction of North Africans to European control compare to their reaction to rule by the Islamic Empire?

_____

_____

_____

_____

Writing

Check for Understanding

1. Informative/Explanatory What are hieroglyphics? How do hieroglyphics help historians know more about ancient Egypt?

_____

_____

_____

2. Informative/Explanatory Write a summary of the events and results of the Arab Spring.

_____

_____

_____

networks

North Africa

Lesson 3: Life in North Africa

ESSENTIAL QUESTION
Why do conflicts develop?

Terms to Know
souk open-air market
fellaheen poor farmers of Egypt
couscous small nuggets of semolina wheat that are steamed
diversified varied
constitution framework of a government

What Do You Know?
In the first column, answer the questions based on what you know before you study. After this lesson, complete the last column.

Now...		Later...
	What religions are practiced in North Africa?	
	What major ethnic groups live in North Africa?	
	What languages are spoken in North Africa?	
	How did the Arab Spring affect people in North Africa?	

North Africa

Lesson 3: Life in North Africa, *continued*

Cultures of North Africa

Guiding Question *What is daily life like in North Africa?*

Three main groups make up the population of North Africa: Egyptians, Berbers, and Arabs. The region has a varied culture. Egypt has an ancient heritage. French influence can be seen in Morocco and Tunisia. Although Arab Muslim culture dominates, some Berber traditions continue.

Most people are Muslims, but some Christians and Jews also live in the region. One in 10 of Egypt's people are Christians. Most are members of the Coptic Christian Church.

Of the North African nations, Libya has the highest rate of urbanization. Only about half of Egypt's people are city dwellers. Cairo is the region's largest city with more than 9.3 million people. It has skyscrapers, parks, and mosques, or Islamic places of worship. An important feature of cities is the **souk**, or open-air market. North Africa's largest cities are shown in the table below.

City	Country
Cairo	Egypt
Algiers	Algeria
Casablanca	Morocco
Tunis	Tunisia

Farming villages in rural Egypt can be as small as 500 people. Families live in homes built of mud brick with few windows. Each morning, poor farmers called **fellaheen** work all day in fields.

Farms in Libya cluster around oases. In Morocco, many farmers build terraces on steep hillsides to plant their crops. Some rural dwellers in North Africa still live like nomads. They tend herds of sheep, goats, or camels, and move from place to place in search of food and water.

In Morocco, Algeria, and Tunisia, the base of many meals is **couscous**, small nuggets of semolina wheat that are steamed. Sandwiches are made with flat pieces of pita bread and grilled meat or fish. People also eat falafel—ground, dried beans that are formed into cakes and fried.

The arts in North Africa reflect the culture of Islam, which forbids art that shows animals or humans. Folk art has intricate patterns but no figures. Many young people are attracted to Western music and movies, but this has angered strict Muslims.

Defining

1. What is a *souk*?

   _____

   _____

   _____

Marking the Text

2. Highlight the different ways people live in rural areas.

Making Connections

3. Why does the folk art of the region feature intricate patterns?

   _____

   _____

   _____

   _____

   _____

Reading Progress Check

4. What is an example of the influence of Islam on daily life?

   _____

   _____

   _____

   _____

   _____

   _____

Reading Essentials and Study Guide **229**

Lesson 3: Life in North Africa, *continued*

Marking the Text

5. Circle three social issues that are concerns in North Africa.

? Explaining

6. Why is illiteracy such a problem for the region's people?

_____

_____

_____

✓ Reading Progress Check

7. What might happen in North Africa if young people grow impatient at slow rates of economic growth? Why?

_____

_____

_____

_____

_____

Marking the Text

8. Read the text on the left. Highlight the two forces that helped bring about the Arab Spring.

Arabic is the official language of all five countries. French is prominent in Morocco, Algeria, and Tunisia. French and English are used in cities, but Berber languages are commonly heard in rural areas across North Africa.

Challenges in North Africa

Guiding Question *What challenges face North Africa?*

North Africa has many pressing economic issues. In Libya, the income gained from selling oil did not reach the great mass of people. When Qaddafi fell from power in 2011, Libyans hoped their lives would improve, but progress was slow. Algeria tried to shift its economy away from oil and natural gas. However, the government keeps tight control on businesses, so companies from other countries are not willing to invest there. Morocco's economy is the most **diversified**, or varied, but poverty and unemployment remain a problem.

In recent years, thousands have left North Africa for Europe, looking for jobs. Morocco, Algeria, and Tunisia have lost the most people. Spain and France are the major destinations.

Social issues are also a concern. High populations in Libya and Egypt lead to crowding, poverty, and poor health care. Also a large share of the population is under the age of 14, especially in Egypt and Libya. These countries' economies will need to develop to help these young people find jobs in the future.

Another social issue is literacy. Libya has the highest literacy rate at 89 percent, but it is much lower in other North African countries. Illiteracy is most serious in Morocco. There, more than 65 percent of men can read and write, but less than 40 percent of women can do so. In the other four countries, literacy is about 20 percent lower for women than men.

North Africa's Future

Guiding Question *How will North Africa address the problems it faces?*

Two political forces are strong in North Africa. One is a push for democracy. Another is the rise of Islamic fundamentalism. The political party of the Muslim Brotherhood gained a majority in Egypt and Morocco, and a large share of seats in Tunisia. These forces helped bring about the Arab Spring of 2010 and 2011. However, conditions across the region are uncertain.

Lesson 3: Life in North Africa, *continued*

In 2012, Egypt began writing a new **constitution**, which is a set of rules for a nation and its government. More power might go to parliament, but it is not clear how well the new government will work. In Algeria, Muslim fundamentalists have long pushed for change. There the conflict was violent and many people died. As in Morocco, the government was able to keep power after the Arab Spring, but it had to promise to create more democracy.

By 2012, Libya's victorious rebels were working on making a new government. They faced the need to rebuild much of the country after the civil war. Also, leaders in eastern Libya said they wanted self-rule, which could lead to continued conflict.

Many Muslims worry about the impact of Western culture on their lands. They think Western entertainment conflicts with Islamic values. They also disagree with Western ideas on women's rights. Women in North Africa have more rights than in other Muslim lands, especially in Tunisia where they can own businesses. In Egypt, Coptic Christians worry about their position because of attacks from Muslim extremists.

Egypt and Morocco have close ties to the United States. That friendship has come under increasing criticism from Muslim fundamentalists. This raises questions about what will happen if Muslim conservatives gain power. Will the new governments reject close ties with the United States? Will Algeria and Libya be less willing to sell oil to the United States? Will these new governments take steps against Israel?

Defining
9. What is a *constitution*?

_____

_____

_____

Reading Progress Check

10. Why were the results of the Arab Spring different in Algeria and Morocco compared with the other countries of the region?

_____

_____

_____

_____

Writing

Check for Understanding

1. Informative/Explanatory How is the relatively young population of this region related to the economic issues there?

_____

_____

2. Informative/Explanatory Why is the political situation in North Africa important to the United States?

_____

_____

_____

networks

East Africa

Lesson 1: Physical Geography of East Africa

ESSENTIAL QUESTION
How does geography influence the way people live?

Terms to Know
rift to separate from one another
desertification the process by which agricultural land is turned into desert
hydroelectric power the production of electricity through the use of falling water
geothermal energy electricity produced using underground heat sources, such as hot springs and steam

Where in the World: East Africa

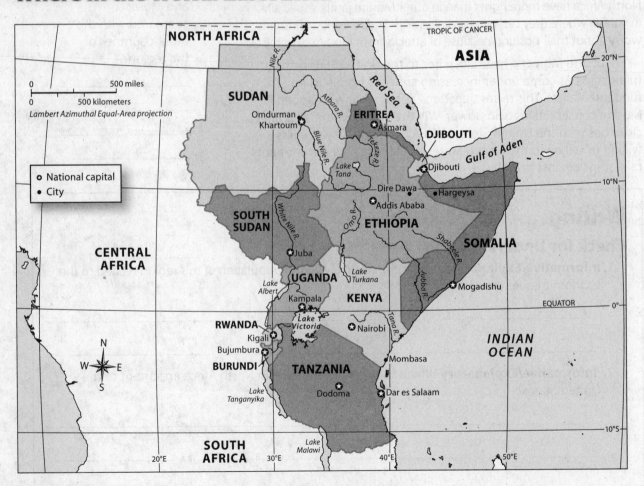

networks

Lesson 1 Physical Geography of East Africa, *continued*

Land and Water Features

Guiding Question *What makes the ecosystem of East Africa diverse?*

The Great Rift Valley is the most unusual feature of East Africa's physical geography. It is actually a series of large valleys and depressions on Earth's surface. The Great Rift started forming about 20 million years ago when tectonic plates began to tear apart from one another. Africa was once connected to the Arabian Peninsula. As the two **rifted** apart, or separated from one another, the land in between sank and was filled by the Red Sea. Eventually, all of East Africa will separate from the rest of Africa, and the Red Sea will fill the rift.

> **Great Rift System**
>
> • About 4,000 miles (6,437 km) long and 30 to 40 miles (48 to 64 km) wide
>
> • Eastern branch starts in Southwest Asia, runs through the Danakil plain in Ethiopia, and takes the form of deep valleys as it extends into Kenya and Tanzania.
>
> • Western branch stretches from Lake Malawi in the south through Uganda in the north.

Much volcanic and seismic activity has occurred along the branches of the Great Rift Valley. The largest volcanoes are located on the eastern Rift. They include Kilimanjaro, between Kenya and Tanzania. Kilimanjaro is the tallest mountain in Africa. Its summit is covered with snow year-round, even though it is near the Equator.

Sudan is home to vast plains and plateaus. In the north, the terrain is desert. Somalia, on the east, is also extremely dry. It is made up largely of savanna and semidesert. Djibouti, to the north, has a very diverse landscape of rugged mountains and desert plains. South of Sudan, the Ruwenzori Mountains divide Uganda from the Democratic Republic of the Congo. Small, landlocked Rwanda is known as "the land of a thousand hills."

The longest river in the world is the Nile (4,132 miles or 6,650 km). The Nile Basin includes parts of many countries in East Africa: Tanzania, Burundi, Rwanda, Kenya, Uganda, Ethiopia, South Sudan, and the Sudan. It has two sets of headwaters, the Blue Nile and the White Nile. The Blue Nile begins in the northern highlands of Ethiopia. The White Nile begins in Lake Victoria and passes through swampy wetlands in central South Sudan. This huge area is called the Sudd.

✏️ Marking the Text

1. Read the text on the left. Underline the sentence describing what will eventually happen along the Great Rift.

❓ Locating

2. Where does each of the two Nile tributaries begin?

   _____

   _____

   _____

❓ Identifying

3. What is the Sudd?

   _____

   _____

   _____

   _____

☑️ Reading Progress Check

4. What caused the striking physical features of the Great Rift Valley in East Africa?

   _____

   _____

   _____

   _____

networks

East Africa

Lesson 1 Physical Geography of East Africa, *continued*

⚙ **Determining Cause and Effect**

5. How are glaciers affected by climate change?

_____

_____

_____

🔤 **Describing**

6. What is the climate like on the tropical grasslands of Kenya?

_____

_____

_____

⚙ **Determining Cause and Effect**

7. What causes *desertification*?

_____

_____

_____

☑ **Reading Progress Check**

8. What generalization can you make about the variations in temperature in East Africa?

_____

_____

_____

The two branches of the Nile meet at Khartoum, in northern Sudan. The Nile then flows north to the Mediterranean Sea. East Africa has few important rivers, partly because of intermittent rainfall and high temperatures in the region.

Many of East Africa's lakes are located near the Great Rift Valley. The largest of these is Lake Victoria. It is relatively shallow, but contains about 200 species of fish. Lake Tanganyika is the longest freshwater lake in the world, and the second deepest. Farther south is Lake Malawi.

Major Lakes of Africa and Their Bordering Countries	
Lake	**Bordered by**
Victoria	Kenya, Tanzania, Uganda
Tanganyika	Tanzania, Democratic Republic of the Congo
Malawi	Tanzania, Malawi, Mozambique

Climates of East Africa

Guiding Question *How does climate vary in East Africa?*

The diverse geography of East Africa is matched by its varied climate. Temperatures in East Africa tend to be warmer toward the coast and cooler in the highlands. Sudan, Djibouti, and Somalia have high temperatures for much of the year. High mountains such as Kilimanjaro and the Ruwenzori Range have had glaciers for thousands of years. Due to climate change, however, these glaciers are melting. Some experts predict that the glaciers of Kilimanjaro will disappear entirely over the next 20 years. The highlands of Kenya and Uganda have a springlike climate year-round.

In many parts of the region, rainfall is seasonal. This is especially true close to the Equator. Wet seasons alternate with dry ones. On the tropical grasslands of Kenya, for example, two rainy seasons occur in most years. These are the "long rains" of April and May and the "short rains" of October and November. The months in between these periods are dry, with little or no rainfall.

Rainfall in East Africa can be unpredictable. Sparse rainfall can result in drought. An urgent issue in the region is **desertification**. This is the process by which agricultural land turns into desert. It occurs when long periods of drought and unwise land use destroy vegetation. During the last half century, desertification has affected much of the Sahel. This area lies between the Sahara and countries to the south, such as Sudan and South Sudan.

netw⚹rks

Lesson 1 Physical Geography of East Africa, *continued*

Resources of East Africa

Guiding Question *Which natural resources are important in East Africa?*

Minerals, energy resources, landscapes, and wildlife are all important resources found in East Africa. There are small gold deposits along the rifts in Kenya, Uganda, and Tanzania. Gemstones such as sapphires and diamonds are found in Tanzania. Rwanda has tin deposits. Ethiopia and Uganda produce lumber. Lake Assal in Djibouti is the world's largest salt reserve, with more than 1 billion tons. It is located at the lowest point in Africa.

Energy resources include coal in Tanzania and petroleum in Uganda, South Sudan, and northwestern Kenya. Sudan has the opportunity to develop **hydroelectric power**, or electricity produced from falling water. Hydroelectric power is already in use in Kenya and Tanzania. Kenya and Djibouti are favorable locations for the development of **geothermal energy**. This type of energy comes from underground heat sources such as hot springs and steam. It is estimated that 30 percent of Kenya's energy needs could be met by geothermal energy by the year 2030. Energy resources and their use are often inconsistent and uneven. Major cities use much of the energy produced but it is often unavailable in rural areas.

East Africa's land and wildlife are also important assets. The breathtaking scenery of the Great Rift Valley is an important tourist resource. In addition, the region is home to the greatest assemblage of wildlife in the world. There are many national parks and wildlife sanctuaries. Perhaps the best-known reserves are in Kenya and Tanzania. These parks are home to lions, leopards, cheetahs, giraffes, zebras, elephants, and dozens of species of antelope. Every year, tourists from all over the world witness the Great Migration. During that event, more than 1 million animals travel hundreds of miles in search of fresh grazing land.

🖉 Marking the Text

9. Highlight the names of countries with mineral resources in pink. Highlight the names of countries with energy resources in yellow.

❓ Describing

10. Why do you think energy resources are often unavailable in rural areas?

_____

_____

☑ Reading Progress Check

11. What two promising alternatives might help improve energy supplies in the East African region?

_____

_____

Writing

Check for Understanding

1. Argument Which natural resources do you think are most important to East Africa? Why?

_____

_____

_____

East Africa

Lesson 2: History of East Africa

ESSENTIAL QUESTION
Why do people trade?

Terms to Know
tribute a regular tax payment demanded from one country by another
imperialism the control of smaller, weaker nations by another nation
genocide the mass murder of people from a particular ethnic group
refugee a person who flees a country because of violence, war, persecution, or disaster

When did it happen?

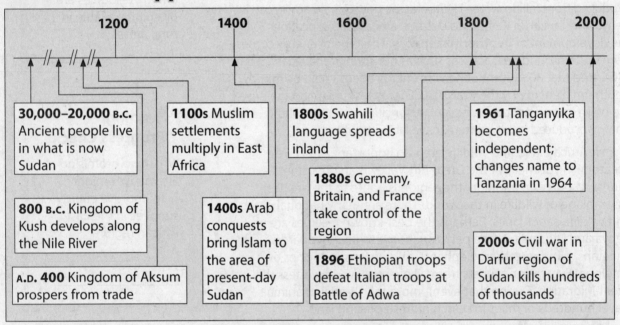

1200	1400	1600	1800	2000

30,000–20,000 B.C. Ancient people live in what is now Sudan

1100s Muslim settlements multiply in East Africa

1800s Swahili language spreads inland

1961 Tanganyika becomes independent; changes name to Tanzania in 1964

800 B.C. Kingdom of Kush develops along the Nile River

1400s Arab conquests bring Islam to the area of present-day Sudan

1880s Germany, Britain, and France take control of the region

A.D. 400 Kingdom of Aksum prospers from trade

1896 Ethiopian troops defeat Italian troops at Battle of Adwa

2000s Civil war in Darfur region of Sudan kills hundreds of thousands

---

Marking the Text

1. Underline the text that discusses how culture was shared between Kush and the Egyptian empire.

Kingdoms and Trading States

Guiding Question *How has the history of trade impacted the region?*

The ancient region of Nubia stretched from the Nile River valley in southern Egypt south almost to what is now Khartoum in Sudan. It extended east to the Red Sea and west to the Libyan Desert. The Nile River was the link between Nubia and the empire of Egypt. In about 1050 B.C., the Kush civilization arose in Nubia. The Kushites traded with the Egyptians and adopted many of their customs. For example, they built pyramids to mark the tombs of their rulers.

East Africa

Lesson 2: History of East Africa, *continued*

Later, they traded with African people south of the Sahara. Around A.D. 350, Kush was conquered by Aksum, a powerful state in what is now northern Ethiopia.

Aksum may have been founded around 1000 B.C. It's location helped it to control the port of Adulis on the Red Sea. At its height, it was the most important trading center in the region. Its connections extended to Alexandria on the Mediterranean Sea. Aksum specialized in sea routes connecting the Red Sea to India. Christianity spread from its origin in Jerusalem along the trade routes. The Aksum kings adopted Christianity as their religion.

Around the A.D. 900s, after the decline of Aksum, Arabs settled on the East African coast of the Indian Ocean. Islam grew steadily more important in the region. The Arabic and Bantu languages mingled to create a new language, Swahili. Swahili is still widely spoken in Tanzania, Kenya, and elsewhere. Gradually, the coastal settlements formed independent trading states.

People from Arabia and Iran founded Kilwa in the late A.D. 900s. It was located on an island just off the southern coast of present-day Tanzania. Its merchants traded for products such as Chinese porcelain and Indian cotton. For two centuries, the city was probably the wealthiest trading center in East Africa.

Early Trading Societies in East Africa		
Society	**Originated**	**Products Sold**
Kush	c. 1050 B.C.	Copper, gold, ivory, ebony, slaves, cattle
Aksum	c. 1000 B.C.	Gold, ivory, glue, candy, gum arabic, raw materials
Kilwa	Late A.D. 900s	Copper, iron, ivory, gold

The Colonial Era

Guiding Question *What was the effect of colonization on East Africa?*

Just before 1500, the Age of Discovery began to affect East Africa. Along with other Europeans, the Portuguese established a sea route to India. From Europe, they sailed around Africa, up the coast of Arabia, and on to India. This was much easier and less expensive than any overland trade routes. In this way, the Portuguese were able to bring back many valuable spices from India.

As trade increased, the Portuguese began to demand **tribute**, or a regular tax payment, from the East African trading kingdoms.

Marking the Text

2. Read the text on the left. Highlight the names of religions that spread to East Africa as a result of trade.

Contrasting

3. How did Kush and Aksum differ from the later Arab trading states?

_____

_____

_____

Reading Progress Check

4. Compare the economies of coastal city-states in East Africa to those of the kingdom of Aksum.

_____

_____

_____

_____

Marking the Text

5. Underline the sentence that explains why the Portuguese preferred sea routes to overland trade routes.

East Africa

Lesson 2: History of East Africa, *continued*

? Identifying

6. Which European nation was the first to gain influence in East Africa?

_____

Aᵇc Defining

7. What were some of the reasons behind European *imperialism* in East Africa?

_____

_____

_____

✓ Reading Progress Check

8. What was the significance of Menilek II's victory at the Battle of Adwa in 1896?

_____

_____

_____

? Identifying

9. How did the British change agriculture in their colony in Kenya?

_____

_____

_____

_____

The Portuguese had religious as well as economic motives. They wanted Christianity to replace Islam as the region's religion. Portuguese influence in the region did not last long. Other countries became interested in colonizing Africa.

In the late 1800s, European nations adopted a policy of **imperialism**. Under imperialism, a stronger nation controls other, weaker nations. The Europeans carved Africa into colonies.

Reasons for Colonization

- Economic profit
- Access to raw materials
- Opening of new markets
- National pride
- Protection of sea routes
- Maintaining the balance of power
- Conversion of Africans to Christianity

Rebellions sometimes challenged colonial rule. One particularly bloody rebellion took place in Sudan, but the British were able to regain control. In Ethiopia, however, the desire for independence prevailed. Italy had colonized neighboring Eritrea. In 1889 the Italians signed a treaty with the Ethiopian emperor, Menilek II. Italy claimed the treaty allowed it to create a "protectorate" in Ethiopia.

Menilek denied Italy's claims. He rejected the treaty in 1893. The Italian governor of Eritrea finally launched a major military attack in response in 1896. Menilek defeated the Italian army at the Battle of Adwa that year. After that, the European powers had to recognize Ethiopia as an independent state. Physical geography played an important role. Rugged mountain terrain made it difficult for attacking forces to invade.

Independence

Guiding Question *How did the countries of East Africa gain their independence?*

After World War II ended in 1945, there was a movement to end colonialism. Weakened European countries granted independence to their East African colonies in the 1960s. Kenya had been a British colony for about 75 years. British plantation owners dominated the economy. They replaced local village agriculture with production of cash crops, such as coffee and tea on a large scale.

networks

East Africa

Lesson 2: History of East Africa, *continued*

Native people were driven off the land. A nationalist named Jomo Kenyatta led a protest movement and negotiated the terms of independence with the British. In 1963 Kenya finally became independent.

Many ethnic groups in former colonies were often in conflict with one another. Rwanda and Burundi are home to two rival ethnic groups. The Hutu are in the majority, and the Tutsi are a minority. In the 1990s, the Hutu-led government of Rwanda killed hundreds of thousands of Tutsi in a campaign of **genocide**—the mass murder of a group of people because of their ethnicity.

Since the 1970s, Somalia has been scarred by civil war. There have been border disputes with Ethiopia. Drought has brought famine to much of the country. In late 1992, the United States led an intervention to restore peace. However, the civil war remained unresolved. The instability, misery, and violence in Somalia have affected neighboring countries. For example, thousands of **refugees** have fled to Kenya for safety.

Africa's newest country emerged as a result of civil war. Sudan won independence in 1956. Most people in Sudan are Muslim. In the southernmost 10 provinces, however, most people follow traditional religions or Christianity. They feared that the central government was trying to impose an Islamic state. Religion was just one issue that led to conflict. Economic issues are also a problem. The southern provinces hold many of the area's petroleum deposits. As a result of the civil war, South Sudan became an independent country in 2011.

✏ Marking the Text

10. Underline the definition of *genocide*.

❓ Illustrating

11. Give an example of how unrest in one country affects neighboring countries.

_____

_____

✓ Reading Progress Check

12. How has civil war played an important part in the recent history of East Africa?

_____

_____

_____

Writing
Check for Understanding

1. Informative/Explanatory What impact has trade had on East Africa throughout history?

_____

_____

2. Informative/Explanatory What was the result of civil war in Sudan?

_____

_____

networks

Lesson 3: Life in East Africa

ESSENTIAL QUESTION
Why does conflict develop?

Terms to Know
population density measure showing the average number of people living within a square mile or a square kilometer
clan a large group of people sharing a common ancestor in the far past
subsistence agriculture a type of farming in which the farmer produces only enough to feed his or her family
oral tradition the process of passing stories by word of mouth from generation to generation
poaching the trapping or killing of protected wild animals in order to make a profit

What Do You Know?
In the first column, answer the questions based on what you know before you study. After this lesson, complete the last column.

Now...		Later...
	How diverse is the population of East Africa?	
	How do most people in East Africa live?	
	What is the importance of oral tradition in East Africa?	
	Why is ecology important in East Africa?	

Marking the Text
1. Underline the cities of East Africa that are found on or near the coast of the Indian Ocean.

The People of East Africa
Guiding Question *What ethnic groups contribute to the diversity of the population?*

The population of East Africa is split between large cities and rural areas. Many large cities are on or near the coast of the Indian Ocean. These include Mogadishu in Somalia, Mombasa in Kenya, and Dar es Salaam in Tanzania. Others developed from important trading centers. These cities include Nairobi, the capital of Kenya, and Addis Ababa, the capital of Ethiopia.

East Africa

Lesson 3: Life in East Africa, *continued*

Of the 11 countries in the region, Ethiopia has the largest population. About 80 million people live there. Djibouti has the smallest population, with about 1 million people. **Population density**, or how many people live in a given geographical area, varies greatly. Overall, Rwanda has the highest population density. Somalia has the lowest.

The populations of Kenya, Tanzania, and Ethiopia are ethnically diverse. Ethnic identity is closely linked to language and also to geography. For example, Kenya has many different ethnic groups.

Ethnic Diversity in Kenya by Location	
Kikuyu, Kamba, Meru, Nyika	Central Rift highlands
Luhya	Lake Victoria basin
Luo	lower parts of western plateau
Masai	south, along border with Tanzania
Samburu, Turkana	arid northwest

Another type of ethnic identity is the **clan**. Clans are large groups of people sharing a common ancestor in the far past. In Somalia, the basic ethnic unit is the clan. In countries with many ethnic groups, building a sense of national identity is often difficult. People often feel a stronger allegiance to their ethnic group than to the country. For example, a Somali might feel a greater attachment to his or her clan than the country of Somalia.

Many local languages are spoken in East Africa. People in Ethiopia, for example, speak about 100 languages. Kenya also has a wide variety of spoken languages. Swahili and English are the official languages used in the Kenya National Assembly and the courts. Swahili is almost universal in Tanzania. Somali is the official language of Somalia. However, many people speak Arabic in the north and Swahili in the south. In Somalia's colleges and universities, it is not uncommon to hear people speaking English or Italian.

Most East Africans are either Christian or Muslim. Many traditional African religions also thrive in the region. About 60 percent of Ethiopians are Christians. The Ethiopian Orthodox Church is one of the world's oldest Christian churches. The Kenyan constitution guarantees freedom of religion. Christianity is practiced by more than two-thirds of Kenya's population. Muslims are an important religious minority there. Tanzania is evenly split among Christianity, Islam, and traditional religions.

Defining

2. What sets *clans* apart from other types of ethnic groups?

_____

_____

_____

Marking the Text

3. Read the text on the left. Highlight the names of languages spoken in East Africa.

Identifying

4. Which religions are practiced by most people in East Africa?

_____

_____

Reading Progress Check

5. In a region with such diverse languages, how do you think East Africans can communicate with people outside their own language group?

_____

_____

_____

_____

East Africa

Lesson 3: Life in East Africa, *continued*

Marking the Text

6. Read the text on the right. Underline a sentence that explains what *subsistence agriculture* is.

Analyzing

7. Why do you think the governments of Tanzania and Kenya want to persuade the Masai to settle in one place?

_____

_____

_____

Defining

8. What are some examples of *oral tradition*?

_____

_____

_____

Reading Progress Check

9. Compare and contrast urban and rural daily life in East Africa.

_____

_____

_____

_____

Life and Culture

Guiding Question *What is daily life like for people in East Africa?*

Most East Africans live in the countryside. Cities are growing rapidly, however, due to the economic opportunities they provide. Nairobi is Kenya's capital and most important industrial city. With more than 3 million people, it is the most populous city in East Africa. It is a city of contrasts. High rise business and apartment buildings sit near slums built of scrap metal.

Daily life in rural areas is very different from life in cities. A rural family might live in a thatched-roof dwelling with no plumbing. Often, there is no electricity available. Some rural people practice **subsistence agriculture**. They grow crops to feed themselves and their families. Other rural people grow cash crops to sell.

One rural group is the Masai, who live in northern Tanzania and southern Kenya.

The Masai
• Nomadic cattle herders
• Governments of Tanzania and Kenya have tried to persuade the Masai to abandon their nomadic life style.
• Masai want to preserve their traditional way of life.

East African culture is deeply influenced by **oral tradition**. This means that stories are passed by word of mouth from one generation to the next. Folktales and fables are good examples of oral tradition. In Kenya, hymns of praise were passed on to support independence.

The leading novelist in East Africa is Kenya's Ngugi wa Thiong'o. Ngugi writes in English, as well as in the Bantu language of the Kikuyu people. In Tanzania, a form of music called *tarab* combines African, Arab, and Indian elements and instruments. It has developed an international following.

East Africa is also important in the fields of anthropology and ecology. Evidence indicates that humans originated in East Africa. The earliest known human bones come from Kenya and Ethiopia. The fossil beds of Olduvai Gorge in northern Tanzania have furnished us with an important record of 2 million years of human evolution. In the domain of ecology, the national park systems of East Africa are without equal. Protected areas like the Serengeti National Park in Tanzania and Volcanoes National Park in Rwanda are preserving a precious inheritance.

East Africa

Lesson 3: Life in East Africa, *continued*

Challenges

Guiding Question *How do economic, environmental, and health issues affect the region today?*

Agriculture is the main economic activity in East Africa. Farmers in the region face difficult challenges. Soils there are not very fertile, and rainfall can be intermittent. Government policies favor the production of cash crops for export.

The region is one of the poorest in the world. The population in many countries is growing at a faster rate than the world's average. Industrialization has come slowly. Economies have suffered because of civil war and political instability. The economy is also linked to the availability of transportation, communication, and education. Literacy rates range from a low of 38 percent in Somalia to 87 percent in Kenya.

East Africa's lack of electric power has accelerated deforestation. People cut down trees to meet their energy needs. They use the wood to cook food and heat their homes. In countries like Sudan, desertification also poses serious problems.

The countries of East Africa have hoped to boost their economies and preserve their heritage by setting up national parks and wildlife sanctuaries. Ecotourism is important to the East African economy. However, wild animals such as elephants and lions face the threat of **poaching**, or illegal hunting. Elephants are killed for their ivory tusks.

Poor nutrition is a major problem in the region. Conflicts in several countries have halted economic development and caused widespread starvation. HIV/AIDS is a serious and often fatal disease. It is a major health issue in Kenya, Tanzania, and Ethiopia. Resources required for medical education and treatment have further strained East African economies. Because of AIDS, life expectancy in the region is about 20 years lower than in the United States.

Analyzing

10. What are two of the most important challenges confronting East Africa today?

_____

_____

Marking the Text

11. Read the text on the left. Underline a sentence that explains how HIV/AIDS affects the economies of East African countries.

Reading Progress Check

12. What is one major cause of deforestation in East Africa?

_____

_____

Writing

Check for Understanding

1. **Informative/Explanatory** What challenges do farmers in East Africa face?

_____

_____

_____

Central Africa

Lesson 1: Physical Geography of Central Africa

ESSENTIAL QUESTION

How do people adapt to their environment?

Terms to Know

watershed the land drained by a river and its system of tributaries

estuary an area where river currents and ocean tide meet

slash-and-burn a method of clearing farmland by cutting down and burning trees and shrubs in order to make the soil more fertile

biodiversity variety of plants and animals living on the planet

Where in the World: Central Africa

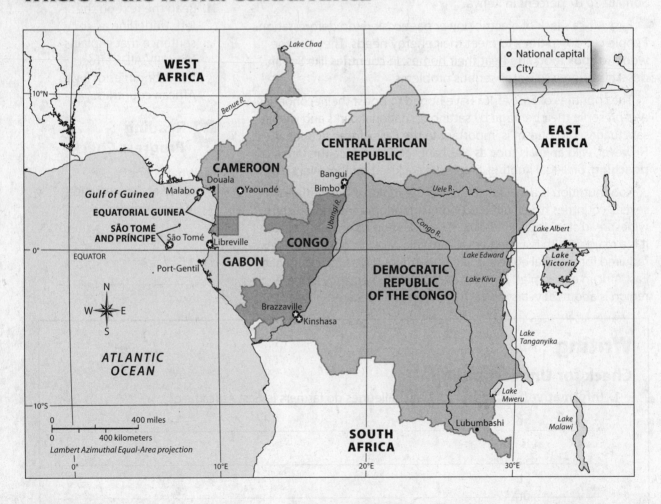

Central Africa

Lesson 1: Physical Geography of Central Africa, *continued*

Landforms and Waterways

Guiding Question *What makes some landforms and waterways so important to the region?*

Central Africa is located along and near the equator. It consists of seven countries, as shown in the table below.

Central Africa
The Democratic Republic of the Congo (DRC)
The Central African Republic
The Republic of the Congo
Cameroon
Gabon
Equatorial Guinea
The island country of São Tomé and Príncipe

The main landform of Central Africa is the **watershed** of the Congo River. A watershed is the land drained by a river and its system of tributaries. At the center of the watershed lies a depression called the Congo Basin. There are high plateaus along most of its sides.

The eastern edge of Central Africa is the Great Rift Valley. Here, rugged mountain ranges soar above a broad, deep valley that holds several long narrow lakes. Along the Atlantic Coast lies a narrow lowland. Several important islands off the coast are part of Central Africa. Two of these form the country of São Tomé and Príncipe. Two other islands, Bioko and Pagalu, belong to Equatorial Guinea. This country also includes several smaller islands and territory on the mainland. The Central African Republic is the only landlocked country in this region.

No other river system in Africa offers as many miles of navigable waterways as the Congo and its tributaries. A navigable river is one on which ships and boats can travel. Although the Congo River is an important waterway, cataracts (waterfalls) and rapids prevent the uninterrupted passage of ships from one end to the other.

Cataracts close to the mouth of the Congo keep seagoing ships from traveling farther inland. Downstream, the river widens into an **estuary**, a passage in which freshwater meets salt water.

The Congo River is important to Central Africa. People use its water for agriculture and depend on its fish for food. The river is also a vital means of transportation. Finally, dams on the river generate large amounts of hydroelectric power.

Marking the Text

1. Circle the islands that belong to Equatorial Guinea. Underline the only landlocked country.

Describing

2. Why can't seagoing ships travel inland very far from the mouth of the Congo?

_____

_____

_____

_____

Defining

3. What is an *estuary*?

_____

_____

_____

_____

Reading Progress Check

4. Give two reasons for the Congo River's importance to Central Africa.

_____

_____

_____

_____

netw⚹rks

Lesson 1: Physical Geography of Central Africa, *continued*

 Explaining

5. How does being on or near the Equator affect the length of days and the weather?

_____

_____

🖉 **Marking the Text**

6. Highlight the text that describes Central Africa's rain forest.

Aᵇc Defining

7. In your own words, describe *slash-and-burn agriculture*.

_____

_____

_____

✓ **Reading Progress Check**

8. Imagine that you are living in northern Cameroon. You decide to move to a location in the Republic of the Congo near the Equator. What changes in climate can you expect?

_____

_____

_____

Climate and Vegetation

Guiding Question *What are the prevailing climates in Central Africa?*

The area of Central Africa that lies along the Equator has a warm, wet climate. Because of its location, this zone experiences little seasonal variation in weather and length of daylight. The midday sun is directly or almost directly overhead every day, and daytime temperatures are always high. Rainfall is abundant throughout the year, with totals greater than 80 inches (203 cm) in some areas.

To the north and the south of the region's equatorial zone lie tropical wet-and-dry climate zones. As the name suggests, these zones have both rainy and dry seasons. There is great climate variation within the zones. In the areas nearest the Equator, the dry season typically lasts about four months. In the areas farthest from the Equator, it might last as long as seven months.

A tropical rain forest, the second largest in the world, covers more than half of Central Africa. In this rain-soaked realm, closely packed trees soar as high as 15-story buildings. The forest also holds a tremendous variety of other plants, some of which are used in traditional medicines.

The interwoven crowns of the trees create a dense canopy, or roof, that blocks nearly all of the sunlight, leaving the lower levels in gloom. Because of the lack of sunlight, the forest floor has few of the small plants found in other types of forests.

Wind damage or death of a tree creates opening in canopy		Tree seedlings and saplings get a chance to grow tall

In northern and southern Central Africa, rain forests give way to savannas. These are areas with a mixture of trees, shrubs, and grasslands. Experts believe that the size of savanna areas may have increased because of **slash-and-burn agriculture**. This is when trees and shrubs are cut down and burned in order to make the soil more fertile. After a few years the soil's fertility becomes low, so farmers move to another area. When trees and shrubs return, the farmers come back and repeat the process. This type of agriculture destroys plant and animal habitat, creates air pollution, and is causing the world's tropical forests to shrink at an alarming rate.

Supporters of environmental conservation worry that activities which harm rainforests threaten **biodiversity**. Biodiversity is the wide variety of life on Earth. They argue that if ecosystems are wiped out, valuable species will be lost forever.

Central Africa

Lesson 1: Physical Geography of Central Africa, *continued*

Natural Resources

Guiding Question *Which natural resources are important in Central Africa?*

The greatest abundance of mineral resources in Central Africa is found in the Democratic Republic of the Congo. The country holds deposits of cobalt, copper, gold, uranium, diamonds, iron ore, and limestone.

Gabon produces more than a tenth of the world's supply of manganese. This hard, silvery metal is used in the manufacture of iron and steel. The country also possesses uranium, diamonds, gold, and high-quality iron ore. Equatorial Guinea has deposits of uranium, gold, iron ore, and manganese.

Some mineral resources in Central African countries remain underdeveloped. This is because of political instability, civil conflict, and the high cost of investment. In addition, the region lacks good transportation networks.

Central Africa is rich in other resources. The DRC has the Congo River for potential hydroelectric power. It also has waterways for fishing as well as forest resources.

Large reserves of petroleum and natural gas have been discovered under the seafloor off Equatorial Guinea's Atlantic Coast. Cameroon has natural gas deposits, but they remain untapped. Nearly all of Cameroon's energy comes from dams that generate hydroelectricity.

? Listing

9. What three non-mineral resources does the DRC have?

_____

_____

_____

✔ Reading Progress Check

10. What are two factors that have slowed development of Central Africa's rich natural resources?

_____

_____

_____

_____

Writing
Check for Understanding

1. **Informative/Explanatory** What conclusion can you make about the prevailing climate conditions in areas near the Equator?

_____

_____

2. **Argument** Why might slash-and-burn agriculture be harmful to a country's land in the long term?

_____

_____

_____

netw⊙rks

Central Africa

Lesson 2: History of Central Africa

ESSENTIAL QUESTION
How does technology change the way people live?

Terms to Know

millet a grass that produces edible seeds

palm oil an oil used in cooking

cassava a tuberous plant that has edible roots

colonialism the political and economic rule of one region or country by another country, usually for profit

missionary person who moves to another area to spread his or her religion

coup a sudden violent overthrow of a government, usually by the military

---

When did it happen?

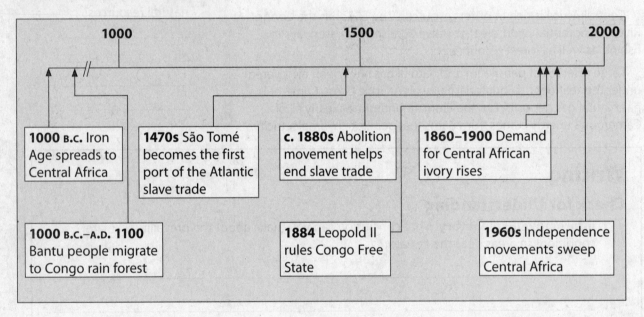

| 1000 | 1500 | 2000 |

1000 B.C. Iron Age spreads to Central Africa

1470s São Tomé becomes the first port of the Atlantic slave trade

c. 1880s Abolition movement helps end slave trade

1860–1900 Demand for Central African ivory rises

1000 B.C.–A.D. 1100 Bantu people migrate to Congo rain forest

1884 Leopold II rules Congo Free State

1960s Independence movements sweep Central Africa

Central Africa

Lesson 2: History of Central Africa, *continued*

Early Settlement

Guiding Question *How did agriculture and trade develop in Central Africa?*

Around 10,000 years ago, Earth's climate entered a dry phase. This meant that the inhabitants of the region had to get more food from a smaller area. Starting in the north, an agricultural revolution moved slowly across the region. People began collecting plants on a more regular basis. They invented specialized digging tools, such as stone hoes, to help with this job. They learned how to clear plots of fertile land and plant a piece of each root or tuber they ate. Over time, hunters and gatherers became farmers.

Gatherers began to collect plants	They invented specialized digging tools	They cleared plots of land to plant roots	Hunters and gatherers became farmers

Cereal farming was the next agricultural development. In the savannas of the north, people began cultivating **millet** and sorghum, two wild grasses that produce edible seeds. Early farmers cultivated trees and gathered their fruit. From the fruit of oil palms they made a cooking oil that was rich in proteins and vitamins. The nutrition boost provided by **palm oil** helped people to become healthier, and improved health brought about population growth.

The increased food supply meant that people could live in larger, more settled communities. This laid the foundation for village life and the development of crafts. In addition, people began using tools of iron instead of stone. They also began using other minerals, such as copper and salt.

European Contact and Afterward

Guiding Question *How did colonization by foreign countries affect Central Africa?*

As European ships came to the region in the 1400s, Central Africa became a busy hub of the slave trade. This was because Europeans wanted a large workforce for their plantations in the Americas.

In the 1400s, Portugal established a colony on the island of São Tomé in order to grow cocoa and sugar. Later this island became a staging area for the transportation of African slaves to the Portuguese colony of Brazil. In Gabon, slaves were gathered in the country's interior and taken to a coastal inlet called the Gabon Estuary, where they were held until European ships arrived.

? Paraphrasing

1. In your own words, describe the agricultural revolution.

_____

_____

_____

_____

Ꭺᵇ꜀ Defining

2. What is *millet*?

_____

_____

_____

✓ Reading Progress Check

3. Why was the agricultural revolution so important for the development of Central Africa as a region?

_____

_____

_____

_____

✏ Marking the Text

4. Circle two Central African nations that were involved in dealing with enslaved people.

Lesson 2: History of Central Africa, *continued*

Marking the Text

5. Draw a diagram of the three stages of triangular trade, showing the products carried in each stage.

Explaining

6. Why were Europeans at first not interested in colonizing Africa?

_____

_____

_____

Reading Progress Check

7. How was Central Africa affected by a conference in Germany in the mid-1880s?

_____

_____

_____

_____

_____

The slave trade was part of what is called the "triangular trade," named for the pattern formed by the trade. In the first stage, ships would sail to Africa carrying goods such as cloth, beads, metal goods, and guns. These goods would be traded for slaves. In the Middle Passage, the ships would carry the slaves to the Americas. There, slaves would be exchanged for goods such as rum, tobacco, molasses, and cotton that were produced on slave-labor plantations. In the third stage, the ships would return to Europe.

European countries brought two important plants back to Europe and Africa from the Americas. These were cassava and maize, which is often called corn. The roots of the **cassava** plant are used to make flour, breads, and tapioca.

In the 1500s and 1600s, European countries began to practice **colonialism**. This is the political and economic rule of one region or country by another country, usually for profit. Europe's first colonies were in the Americas. Central Africa's rugged terrain and tropical diseases made Europeans less interested in colonizing there.

In 1884–1885, Germany hosted a conference of European countries. The countries agreed on a plan to divide Africa into colonies that could be used for European profit. King Leopold of Belgium took over an area that became the Congo Free State. He wanted to make money on the area's rubber plants. Other countries colonized Central Africa, as shown in the table below.

European Country	Central African Colony
Belgium	Belgian Congo
France	Republic of the Congo, Gabon, Central African Republic
Spain	Equatorial Guinea
Germany	Cameroon
Portugal	São Tomé and Príncipe

Europeans justified their economic exploitation of Africa in several ways. They claimed they wanted to promote civilization and to spread Christianity. Europeans sent many **missionaries** to Africa in order to convert the native people. At the same time, Europeans often treated African workers harshly. By the early 1900s, small revolts against French rule were common. People in the Congo Free State suffered severe hardships under King Leopold. Pressured by growing outrage, the Belgian parliament took over the vast area from King Leopold.

Lesson 2: History of Central Africa, *continued*

Independent Countries

Guiding Question *What effects did gaining independence have on the countries of Central Africa?*

In 1960, four French colonies became independent: Gabon, the Republic of the Congo, the Central African Republic, and Cameroon. In the same year, the Democratic Republic of the Congo won independence from Belgium.

Many of these new countries experienced hard times after independence. Their people suffered through periods of ethnic conflict, harsh rule, and human rights abuses. In the Central African Republic, military officer Jean-Bédel Bokassa staged a **coup**. A coup is when a group of individuals seize control of a government. While in power, Bokassa ruled as a dictator and called himself emperor.

There have also been success stories, such as Gabon. Thanks in large part to its plentiful natural resources, Gabon has become one of the wealthiest and most stable countries in Africa.

Equatorial Guinea won independence from Spain in 1968. The first president assumed power in 1973 but was overthrown by his nephew in 1979. Portugal granted independence to São Tomé and Príncipe in 1975. The country has experienced instability and corruption since independence.

? Explaining

8. Why was Gabon able to achieve some stability?

✓ Reading Progress Check

9. Since the countries of Central Africa gained independence, name at least two factors that have stood in the way of political and economic progress.

_____

_____

Writing

Check for Understanding

1. **Informative/Explanatory** What motivated Europeans to colonize Central Africa? How did they justify their colonization?

_____

_____

2. **Narrative** Write a paragraph explaining the development of the slave trade in Central Africa.

_____

_____

_____

networks

Central Africa

Lesson 3: Life in Central Africa

ESSENTIAL QUESTION
Why do people make economic choices?

Terms to Know

refugee person who flees to another country to escape persecution or disaster

trade language common language used to conduct trade

What Do You Know?

In the first column, answer the questions based on what you know before you study. After this lesson, complete the last column.

Now...		Later...
	Where do the people of Central Africa live?	
	What kinds of food are eaten in Central Africa?	
	What conflicts are caused by developing industries in Central Africa?	

Central Africa

Lesson 3: Life in Central Africa, *continued*

The People of Central Africa

Guiding Question *What are some of the differences found among the people of Central Africa?*

Central Africa is home to around 105 million people. The Democratic Republic of the Congo is the region's most populous country. Compared to other parts of the world, Central Africa does not have a large population or a high population density. However, it has a high population growth rate. The median age is below 20, which means that children and teenagers make up about half of the population.

Life expectancy varies greatly. For instance, in the Central African Republic, life expectancy is about 50 years. In Equatorial Guinea and São Tomé and Príncipe, it is around 63.

Hundreds of ethnic groups live in the region. Cameroon and the DRC each have more than 200 groups. Some groups live in two or more countries. One group called the Fang live in southern Cameroon, mainland Equatorial Guinea, and northern Gabon. Another group, the Bambuti (sometimes called Mbuti) dwells in forested areas of the DRC.

Central Africa's population also has many **refugees**. These are displaced people who have been forced to leave their homes because of war or injustice. For example, there was a brutal civil war in the Democratic Republic of the Congo between 1997 and 2003. Thousands of refugees fled the conflict.

Most of Central Africa's people live in rural areas and make their living through subsistence farming. This means growing just enough food to feed the farmer's family. However, there are also large urban areas, and the ratio of city-dwellers to rural residents is increasing rapidly. Many of the chief cities are capitals of countries. In a few countries, such as Gabon and the Central African Republic, more than half the people live in cities.

Hundreds of different languages are spoken across Central Africa. For example, the languages that are spoken in Cameroon are shown in the table below.

Area of Cameroon	Language Spoken
North	Sudanic family
South	Bantu
West	Semi-Bantu

 Activating Prior Knowledge

1. Explain what it means when a country has a high population density.

_____

_____

_____

_____

? Explaining

2. How much longer on average do people in Equitorial Guinea live than people in the Central African Republic?

_____

✎ Marking the Text

3. Highlight the text that discusses two ethnic groups in Cameroon and the DRC.

🔤 Defining

4. What are *refugees*?

_____

_____

_____

Central Africa

Lesson 3: Life in Central Africa, *continued*

Copyright © The McGraw-Hill Companies, Inc.

During the colonial era, people needed a common language, especially when they conducted trade. Certain **trade languages** emerged, and French became the most common trade language of Central Africa.

Central Africa also has many religions. For instance, in the DRC, about 50 percent of the people are Roman Catholic. The rest are Protestant, Muslim, belong to a local sect called the Kimbanguist Church, or follow traditional African religions.

How People Live

Guiding Question *What are some of the key aspects of daily life and culture in Central Africa?*

Daily life in Central Africa is a blend of the traditional and modern. In the countryside, most people practice subsistence farming. They grow crops and raise livestock on small plots of land. If they raise more than they use, they can sell the surplus for cash.

Houses are built using materials such as mud, sundried mud bricks, wood, bark, and cement. Others are wattle-and-daub houses. In these houses, vertical poles driven into the ground are woven with slender, flexible branches to create the wattle. The finished wattle is plastered with mud or clay, known as daub.

In many rural areas, women gather, produce, and prepare the food. The men hunt, trap, or fish for food. They also grow crops such as coffee, cotton, and tobacco for sale. Central African people eat okra, cassava, squash, rice, game, fish, and peanuts.

Many food-related customs of Central Africa may seem surprising. In some areas, for example, males eat in one room while females eat in another. Generosity and hospitality are so deeply ingrained in some cultures that hosts might offer guests an abundance of food while remaining hungry themselves.

The arts are important to the cultures of many groups. The Yaka create decorated masks and figurines. The Kongo people produce statues embedded with pieces of metal. The Luba people create carvings that **depict** women and motherhood. The Mangbetu people are known around the world for their sculpture and pottery. The Fulani people decorate leather items and gourds with elaborate designs.

The region has produced well-known novelists, playwrights, and poets. The writers of Equatorial Guinea, formerly a colony of Spain, often write in Spanish. In music, African jazz called OK jazz, originated in the DRC. Drumming and flute music are also popular.

✓ Reading Progress Check

5. Where do most of Central Africa's people live? How do they make their living?

_____

_____

_____

_____

✎ Marking the Text

6. Underline the different kinds of houses found in villages.

? Explaining

7. Name two duties that women perform in rural areas. Name two duties that men perform.

_____

_____

_____

_____

✓ Reading Progress Check

8. What is subsistence farming?

_____

_____

_____

network **s**

Central Africa

Lesson 3: Life in Central Africa, *continued*

Regional Issues

Guiding Question *What are the greatest challenges confronting Central Africa?*

Central African countries depend heavily on the economic activities shown in the diagram below.

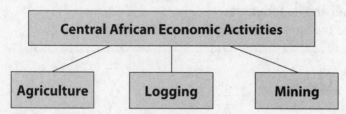

As their economies grow, the region's countries use more resources. They also produce more pollution and waste. In addition, activities such as mining and logging take place in areas with high biodiversity. Critics fear that biodiversity is being lost.

The result is tension between the forces of economic growth and the forces of environmental conservation. In some areas, these tensions have at times led to conflict between local people and outside groups. Development in Central Africa also raises other questions. Should the profits from economic activities go to foreign corporations and investors, or should they remain with the national governments?

✏️ Marking the Text

9. Highlight the two main groups feeling tension because of economic growth.

✅ Reading Progress Check

10. Why are some people critical of mining and lumbering activities in the region?

_____

_____

_____

_____

Writing

Check for Understanding

1. **Informative/Explanatory** Why was the use of trade languages helpful in Central Africa?

_____

_____

2. **Informative/Explanatory** Write a short paragraph explaining some implications of economic development in the countries of Central Africa.

_____

_____

_____

_____

West Africa

Lesson 1: Physical Geography of West Africa

ESSENTIAL QUESTION

How does physical geography influence the way people live?

Terms to Know

basin an area of land that is drained by a river and its tributaries

harmattan a hot, dry wind that blows across the Sahara, carrying large amounts of dust

Where in the World: West Africa

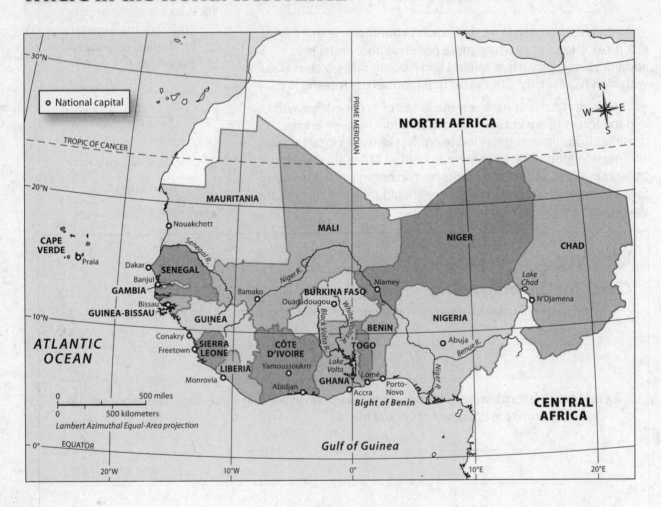

networks

West Africa

Lesson 1: Physical Geography of West Africa, *continued*

Landforms and Bodies of Water

Guiding Question *In what ways do major rivers contribute to the economies of West African countries?*

Erosion has worn most of the land of West Africa into a rolling plateau that slopes to sea level on the coasts. The region has no major mountain ranges. It does have highland regions and isolated mountains. The Air Massif is a group of mountains in central Niger. The Tuareg people graze their livestock in the fertile valleys between its mountains. The highest elevations in West Africa are in the Tibesti Mountains, mostly in northwest Chad.

Southeast of the Tibesti Mountains is the Ennedi, a desert plateau region. The Jos plateau in central Nigeria is mostly open grassland and farmland. In central Guinea is a highland region of savanna and deciduous forest known as the Fouta Djallon. To the southeast it becomes the Guinea highlands, a humid, densely forested region.

The Niger River is West Africa's longest and most important river. It begins in the Guinea highlands and flows northeast toward the Sahara. In Mali, it enters an "inland delta." There it spreads out across the relatively flat land. During the rainy season, the area floods completely. In the dry season, the waters recede, leaving fertile farmland. After it flows past Timbuktu, the river again becomes a single channel.

The Benue River joins the Niger as it flows through Niger and Nigeria, doubling the Niger's volume of water. Where it reaches the Gulf of Guinea, it forms the Niger delta. This is a great **basin**, a depressed area of land drained by a river and its tributaries. Along its course, the Niger provides water for irrigation and hydroelectric power. It is the main source of Mali's fishing industry. It also serves as an important route for transporting goods.

The Senegal River rises in Guinea, then flows northwest to the Atlantic Ocean. It marks the border between the countries of Mauritania and Senegal.

The Black Volta River and the White Volta River originate in Burkina Faso and flow into Ghana. The Akosombo Dam is located where the two rivers once met. It forms one of the world's largest artificial lakes, Lake Volta. The dam provides most of Ghana's electrical needs, and the lake provides water for irrigation.

Lake Chad covers portions of Niger, Nigeria, Chad, and Cameroon. A series of droughts caused Lake Chad to shrink. Another cause was the amount of water taken from the lake and from the rivers that feed it to irrigate crops.

✏️ Marking the Text

1. Read the text on the left. Underline the sentence that tells what the Air Massif is.

🔤 Defining

2. What is the definition of *basin* as it applies to West Africa's geography?

_____
_____
_____
_____

❓ Explaining

3. Why has Lake Chad changed in size over time?

_____
_____
_____
_____

☑️ Reading Progress Check

4. Describe the landforms of West Africa.

_____
_____
_____
_____
_____

networks

West Africa

Lesson 1: Physical Geography of West Africa, *continued*

Climate

Guiding Question *How do amounts of rainfall differ throughout West Africa?*

The climate of West Africa is diverse. It ranges from the harsh, arid Sahara in the north to lush coastal rain forests in the south. In between are vast stretches of grassland—the semiarid Sahel and the rainier savanna. The key feature of the region's climates is its two distinct seasons—the wet season and the dry season.

The Sahel is a semiarid region that runs between the arid Sahara and the savanna. It has a short rainy season. Annual rainfall ranges from 8 to 20 inches (20 to 51 cm). This climate supports low grasses, thorny shrubs, and a few trees. The grasses can support grazing livestock, such as cattle, sheep, camels, and pack oxen. Too many animals leads to overgrazing and permanent damage to the grasslands. Overgrazing and too much farming result in desertification, the process in which semiarid lands become drier and more desertlike.

North of the Sahel is the Sahara. During the day, temperatures are often more than 100°F (38°C) in summer. They can drop by as much as 50°F at night. The few shrubs and small plants that live in the Sahara must go for long periods without water. Some plants send out long roots to water sources deep underground. After a rainfall, plants will suddenly flower. From late November until mid-March, the **harmattan** blows through the Sahara.

The Harmattan
• Intense hot, dry wind that blows through the Sahara
• Carries thick clouds of dust that can extend hundreds of miles over the Atlantic Ocean
• Contributes to the process of desertification

South of the Sahel, the savanna has lusher plant life. The rainy season extends from April to September. The rest of the year is dry. Annual rainfall is between 31 and 59 inches (79 and 150 cm). In some places, though, it can be as little as 20 inches (51 cm). The varying amounts of rainfall make human activity difficult.

Farther to the south, tropical rain forests cover the land. The rain forests receive plenty of rain. They are known for their broad-leaved evergreen trees and rich biodiversity. Rain forests in Sierra Leone and Liberia can receive as much as 200 inches (508 cm) of rain a year.

Marking the Text

5. Read the text on the right. Underline the sentences that describe vegetation in the Sahara.

? Listing

6. List three things that contribute to the process of desertification in West Africa.

_____

_____

_____

A♭c Defining

7. How can the *harmattan* winds influence climate and vegetation?

_____

_____

_____

☑ Reading Progress Check

8. How do humans survive in the Sahel?

_____

_____

_____

_____

West Africa

Lesson 1: Physical Geography of West Africa, *continued*

Resources

Guiding Question *How do West African countries provide for their energy needs?*

West Africa contains many resources. However, some countries lack money to develop them into industries. Nigeria is the region's biggest producer of petroleum and also has large natural gas deposits. Chad has oil fields north of Lake Chad. Benin and Ghana have offshore oil fields. Ghana's oil was considered too expensive to exploit until increased oil prices made drilling profitable.

Ghana's main sources of electricity are two dams on the Volta River. The Senegal River provides about half the energy used in Mauritania. Togo and Nigeria also depend on hydroelectric power.

There are many other important mineral resources in the region.

Mineral Resources of West Africa	
Benin	iron ore, limestone, chromium, gold, marble
Burkina Faso	gold, manganese
Ghana	gold, diamonds, manganese, bauxite, limestone, iron ore
Mali	gold, salt, limestone, iron ore, manganese
Mauritania	copper, iron ore
Nigeria	gold, iron ore, tin, limestone, other minerals
Togo	phosphate, limestone, gold

Many of Mauritania's iron ore deposits have been depleted. Benin is a leader in the production of hardwoods. However, most of the rain forests where this wood comes from have been cleared.

? Identifying

9. What are the most significant energy resources in West Africa, and where are they found?

_____

_____

_____

_____

✓ Reading Progress Check

10. Why is Benin's hardwood industry at risk? What is another industry in this region that has faced a similar problem?

_____

_____

_____

_____

_____

Writing

Check for Understanding

1. Informative/Explanatory Why is desertification an issue in West Africa?

_____

2. Informative/Explanatory Why have some West African countries been slow to develop their resources?

_____

West Africa

Lesson 2: The History of West Africa

ESSENTIAL QUESTION

How do new ideas change the way people live?

Terms to Know
imperialism a policy of seizing political control of other areas to create an empire
secede withdraw

When did it happen?

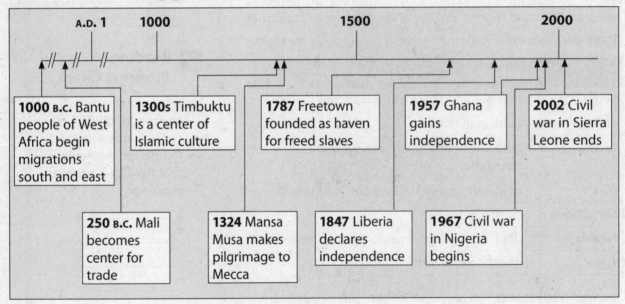

A.D. 1	1000		1500	2000

1000 B.C. Bantu people of West Africa begin migrations south and east

1300s Timbuktu is a center of Islamic culture

1787 Freetown founded as haven for freed slaves

1957 Ghana gains independence

2002 Civil war in Sierra Leone ends

250 B.C. Mali becomes center for trade

1324 Mansa Musa makes pilgrimage to Mecca

1847 Liberia declares independence

1967 Civil war in Nigeria begins

West Africa

Lesson 2: The History of West Africa, *continued*

Ancient Times

Guiding Question *What opportunities did Muslims from North Africa see in West Africa?*

Ten thousand years ago, the Sahara looked more like a savanna than a desert. There were lakes, forests, and large animals including ostriches, giraffes, elephants, antelope, and rhinoceroses. Seminomadic people herded cattle and hunted wild animals. As the climate grew drier, the land supported fewer species of plants and animals. Many people moved to the grasslands to the south. People found that camels can survive without water for long periods. Camels also can carry heavy loads for long distances. They were perfect domesticated animals for desert dwellers.

The Bantu people inhabited West Africa in ancient times. They developed farming as early as 2000 B.C. A vast migration of Bantu people out of West Africa began around 1000 B.C. Their farming practices allowed them to expand into areas occupied by hunter-gather groups. They became dominant in most of East and South Africa. Wherever they moved, the Bantu spread their culture.

Elements of Bantu Culture

• Cultivated bananas, taro, and yams, which thrived in tropical rain forests

• Spread their languages—today, more than 500 Bantu languages are spoken by 85 million Africans

• Had iron-smelting technology for tools and weapons

For thousands of years, the Sahara was a barrier between West Africa and North Africa. By the A.D. 700s, Arabs had conquered most of North Africa. They controlled trade in the region, including Saharan trade routes into West Africa. They soon realized that the region offered opportunities beyond trade, including adding converts to Islam. The Berber people of North Africa was one such people. Groups of Islamic Berbers joined together to become the Almoravids, fierce fighters that wanted to spread their faith.

The Ghana Empire was a powerful West African kingdom. It controlled the gold trade in the region. It traded gold for salt that Arab traders brought in from the Sahara. In the A.D. 1000s, the Almoravids conquered Ghana. Their rule was short, but long enough to damage the trade that kept the empire alive. The dry climate could not support the increase in agriculture necessary to feed the population that had grown with the Almoravids. This resulted in desertification. Ghana became weak, and its conquered neighbors began to break away.

Marking the Text

1. Read the text on the left. Highlight the names of wild animals that once roamed the Sahara.

Determining Central Ideas

2. How did the Bantu culture influence the parts of Africa to which the Bantu migrated?

_____

_____

_____

Identifying

3. Who were the Almoravids?

_____

_____

_____

_____

Reading Progress Check

4. Why did the Ghana Empire collapse?

_____

_____

_____

West Africa

Lesson 2: The History of West Africa, continued

Marking the Text

5. Read the text on the right. Underline the sentences that explain why Timbuktu was important.

? **Explaining**

6. How did the Islamic world become aware of the wealth of the Mali Empire?

_____

_____

⚙ **Contrasting**

7. What distinguished the coastal trading kingdoms from the earlier trading empires of West Africa?

_____

_____

_____

_____

✓ **Reading Progress Check**

8. How did the slave trade in West Africa change after the arrival of the Europeans?

_____

_____

_____

West African Kingdoms

Guiding Question *Why was the city of Timbuktu important to different trading kingdoms in West Africa?*

After Ghana's decline, Mali grew rich from the gold-for-salt trade. It reached its height under the emperor Mansa Musa. The city of Timbuktu was already an important trading post. It also became a center for Islamic culture. Mansa Musa went on a historic pilgrimage to Mecca. By the time he returned home, the Islamic world knew there was a powerful Islamic kingdom south of the Sahara.

After Mansa Musa's death, Songhai replaced Mali as the most powerful West African empire. A well-trained army and a navy that patrolled the Niger River made Songhai the largest of the three trading empires. It controlled the region's trade routes, had salt mines in the Sahara, and sought to become a center of Islamic learning. The empire fell to the Moroccans by the end of the 1500s.

Slavery had been practiced in Africa for centuries. Muslims from North Africa and Asia had bought enslaved people from south of the Sahara. European colonists began to buy slaves to do the work in their colonies in the Western Hemisphere. Small African kingdoms along the Atlantic coast became trading partners with Portugal, Spain, and Great Britain. The slave trade became highly profitable in these kingdoms. When Europeans outlawed the slave trade, the kingdoms' economies began to fail.

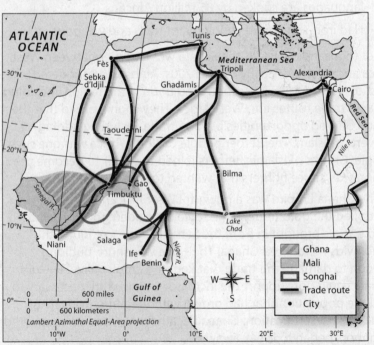

Lesson 2: The History of West Africa, *continued*

European Domination

Guiding Question *What were some of the factors that aroused European interest in exploring and colonizing Africa?*

In 1807 Great Britain outlawed the slave trade. The British had founded the colony of Sierra Leone in 1787 as a safe haven for runaway or freed enslaved persons. In 1822 the Americans founded Liberia as a home to freed American slaves.

To make up for the loss of revenue from the slave trade, the British developed the palm oil trade. As the British started to profit from West African trade, France took notice. In 1869 the Suez Canal opened and diamonds were discovered in South Africa. The two events increased Europe's interest in Africa.

Europeans followed a policy of **imperialism**, the practice of seizing control of other places to create an empire. Great Britain, France, Portugal, Belgium, and Germany met at the Berlin Conference in 1884–1885 to partition Africa.

The French and British claimed most of the territory in West Africa. The table below shows which modern countries in West Africa were formerly French and British colonies.

French and British Colonies in West Africa	
France	Benin, Burkina Faso, Chad, Côte d'Ivoire, Guinea, Mali, Mauritania, Niger, Senegal
Britain	Gambia, Ghana, Nigeria, Sierra Leone, St. Helena

Some colonies were ruled more harshly than others. In all colonies, however, Europeans made the important decisions about how Africans lived. Settlers could force Africans off the most fertile land. Africans resisted, sometimes violently.

New Countries

Guiding Question *How does a ruler gain and hold political power?*

The Gold Coast had been important to European trade since 1487, when Portugal set up its first fort there. By the early 1800s, it was under British control, and in 1874 it became a British colony. Its most important industries were gold, forest resources, and cocoa beans. By the 1920s, the Gold Coast supplied more than half the world's cocoa.

In 1957 the Gold Coast gained independence and became the Republic of Ghana. For many years, it was troubled by conflict. In 1992 Ghana approved a new constitution. Since then, Ghana has become a model of political reform in West Africa.

? Explaining

9. Why did the British encourage development of the palm oil trade?

_____

_____

_____

✏ Marking the Text

10. Underline the text that explains why interest in Africa increased in the mid-1800s.

✓ Reading Progress Check

11. What happened at the Berlin Conference?

_____

_____

_____

? Sequencing

12. Create a timeline of events in Ghana's development as an independent nation.

_____

_____

_____

_____

_____

West Africa

Lesson 2: The History of West Africa, *continued*

⚙ Analyzing

13. Why did Nigeria face so many challenges when becoming an independent country?

_____
_____
_____
_____

✓ Reading Progress Check

14. What are two reasons the French might have had for letting their African colonies declare independence?

_____
_____
_____
_____

Throughout the 1950s, France made concessions to its West African colonies. Many of them instituted some sort of self-rule. Ghana's independence excited and inspired Africans, however. In 1960 Mauritania, Niger, Côte d'Ivoire, Gambia, and Burkina Faso declared their independence. By 1961, all of France's colonies in West Africa were independent.

In 1914 the British combined two colonies to form the Nigerian Protectorate. Several ethnic groups lived in this new territory. When Nigeria gained independence in 1960, tensions flared between them. Because Britain had concentrated most of its development in the south, that is where most well-educated Nigerians lived. They became leaders of the new government. Many of them were Igbo, who were mostly Christian.

When they tried to govern the Muslim Hausa in the north, they were met with hostility. Thousands of Igbo in northern Nigeria were massacred. As many as a million more fled to a region dominated by the Igbo. In 1967 the eastern region **seceded**, or withdrew formally. It became the independent republic of Biafra. Nigerian forces invaded, and after two years of war, Biafra was in ruins. Starvation and disease may have killed more than 1 million people. For long periods since then, the military has controlled Nigeria's government. A new constitution was written in 1978. It was not until 1999, however, that a democratically elected president was able to rule Nigeria.

Military coups and ethnic conflict have also plagued Liberia, spilling over into neighboring Sierra Leone. Estimates are that 50,000 died and another 2 million people lost their homes in the civil war.

Writing

Check for Understanding

1. Informative/Explanatory What opportunities did Muslims from North Africa see in West Africa?

_____
_____

2. Informative/Explanatory Why did European nations want to explore and colonize Africa?

_____
_____

networks

West Africa

Lesson 3: Life in West Africa

ESSENTIAL QUESTION
What makes a culture unique?

Terms to Know

pidgin a simplified language formed by combining parts of several different languages
creole a blend of languages that becomes the language of a region
animist a person who believes in spirits, such as the spirits of ancestors, the air, and the earth
extended family a family made up of several generations, including grandparents, parents, and children
nuclear family the family group that includes only parents and their children
kente colorful, handwoven cloth produced in Ghana
infrastructure a system of roads and railroads that allow the transport of materials

What Do You Know?

In the first column, answer the questions based on what you know before you study. After this lesson, complete the last column.

Now...		Later...
	Why are most West African countries ethnically diverse?	
	What religions are practiced in West Africa?	
	How do West Africans keep their traditions alive?	
	What major challenges do the people of West Africa face?	

The People of the Region

Guiding Question *Why are two or more languages spoken in some West African countries?*

West Africa is the most populous region of Africa. Many different ethnic groups live there. When European countries carved Africa up into colonies to add to their empires, they did not consider the borders of ethnic groups. As a result, European colonies were ethnically diverse, with many different groups living in a colony.

Marking the Text

1. In the text, underline the sentence that explains how different ethnic groups came to live in the same colony.

West Africa

Lesson 3: Life in West Africa, *continued*

? Describing

2. Why are two or more languages spoken in some West African countries?

_____

_____

_____

_____

Aᵇc Defining

3. What are the advantages of a *pidgin* language?

_____

_____

_____

_____

 Marking the Text

4. Underline the sentence that explains the beliefs of *animists*.

✓ Reading Progress Check

5. What are two ways traditions have survived in West Africa?

_____

_____

_____

_____

When the former colonies became independent, the new borders followed the old colonial borders. Many people have a stronger identification with their ethnic group than with the country in which they live. In addition, members of one ethnic group may live in several different countries.

As West African countries gained independence, most held on to the European language that had been used in business and government during the colonial period. Those became the official languages. Sometimes Arabic is also an official language. Most ethnic groups retain their traditional languages. In some places, European and African languages have mixed into a **pidgin** language. This is a simplified language used by people who cannot speak each other's languages but need to communicate. Sometimes two or more languages blend so well that the mixture becomes the language of the region. This is a **creole** language.

Islam was well established in many parts of West Africa long before Christian missionaries arrived from Europe. A large portion of West Africans are still Muslims. In some countries, Christianity is just as dominant, or more so. Many people still practice traditional African religions. Often, they are **animists** and believe in a supreme creator god. Animists believe in spirits--spirits of their ancestors, the air, the earth, and rivers.

The most common settlements in West Africa have traditionally been scattered villages. They are the homesteads of **extended families**, or families made up of parents, children, and other close relatives. The size of the village and its population density depend on how many people the land can sustain. In the north, populations are small and spread out because water is an issue. Most large cities are capital cities. The largest city in West Africa is Lagos, Nigeria, with an estimated population of 10.5 million. It was Nigeria's capital until 1991, when Abuja became the capital.

Life and Culture in the Region

Guiding Question *Why are traditions more important in rural areas of West Africa than they are in West African cities?*

A variety of cultures thrive in West Africa. Countries in the north, such as Mali, Mauritania, and Niger, are more influenced by the cultures of North Africa than are their neighbors. People in cities are more likely to be influenced by Western culture. Far from the city, many people retain their traditional cultures. Village life revolves around the extended family. In cities, the **nuclear family**—parents with their children—is the more common family group.

networks

West Africa

Lesson 3: Life in West Africa, *continued*

City dwellers are more likely to deal with a wide variety of people than are rural dwellers. Capital cities teem with people from different ethnic groups, races, and countries. They are more likely to speak the official language because that is the language they have in common. Ethnicity and tradition are still important in rural areas. Traditional values keep those languages and cultures alive. However, ethnic pride sometimes results in conflicts with neighboring ethnic groups.

West Africans have created unique and important works of art in many fields. Traditional artwork is world famous.

Traditional West African Arts
• Carved masks from Nigeria and Sierra Leone
• Colorful, handwoven cloth called **kente** from Ghana
• Musical stories told by griots blend history and spiritual advice
• Dance celebrates skills and accomplishments

Dance is used for its healing qualities, but people also dance to popular music in clubs. West African music blends many sounds and combines traditional and modern instruments. The Arabic influences of North African music blend with sub-Saharan music, as well as American and European rock and pop music.

West Africa has produced some of the world's finest writers. The most widely read African novel is *Things Fall Apart* by Chinua Achebe of Nigeria. Senegal's first president was a famous poet and lecturer on African history and culture.

Challenges Facing the Region

Guiding Question *How did West African countries build up so much debt?*

Many West African countries face serious problems. The European powers that colonized Africa built the colonies' economies on a few resources, such as gold or peanuts. A colony's **infrastructure**, or underlying framework, was built around those resources. After independence, it became important for the new countries to develop a variety of industries. Otherwise, a drop in the price of a key resource could greatly affect the country's economy. Foreign governments invested in the industrial base of those countries. Often, money was not used wisely. As a result, countries had to deal with enormous debt as well as struggling economies.

❓ Identifying

6. What is one potential downside to preserving traditional values in rural West Africa?

_____

_____

_____

✔️ Reading Progress Check

7. Select an artistic field that you think best shows the culture of West Africa, and explain why you believe it does.

_____

_____

_____

_____

_____

🔤 Defining

8. How did the *infrastructure* of West African colonies lead to poor economies when they became independent countries?

_____

_____

_____

_____

Lesson 3: Life in West Africa, *continued*

? Identifying

9. In your opinion, what is the biggest challenge facing West Africa? Why?

_____

_____

_____

☑ Reading Progress Check

10. Why is education such an important issue for the future of West Africa?

_____

_____

_____

The International Monetary Fund and World Bank do not require poor countries to repay debts they cannot manage. Debt relief has been important for countries such as Ghana, which suffered from bad economic choices and instability. It is now a model for economic and political reform in West Africa. Nigeria also benefited from debt relief and was able to repay what it owed.

Sub-Saharan Africa holds 22.9 million of the 34 million people in the world living with the HIV virus. Dealing with such a large number of HIV-infected people presents several challenges.

Challenges to Fighting HIV/AIDS in West Africa

• Supplying health care to people who carry the virus

• Reducing the number of new infections

• Dealing with people and communities hurt by AIDS-related deaths

---

West Africa's population is growing quickly. To have a sound future, the educational systems must effectively prepare young people for economic and social development. A lack of funding for education means that some countries cannot afford to make updates or improvements to their schools.

Writing
Check for Understanding

1. **Informative/Explanatory** List three ways in which life in cities and life in rural villages differ.

_____

_____

_____

2. **Informative/Explanatory** What are three major challenges to fighting AIDS in West Africa?

_____

_____

_____

Southern Africa

Lesson 1: Physical Geography of Southern Africa

ESSENTIAL QUESTION

How does geography influence the way people live?

Terms to Know

escarpment a steep cliff between a higher and lower surface

landlocked having no border with an ocean or a sea

reservoir an artificial lake created by a dam

blood diamond a diamond that is mined to pay for rebellions and other violent conflicts

poaching illegally killing game

Where in the World: Southern Africa

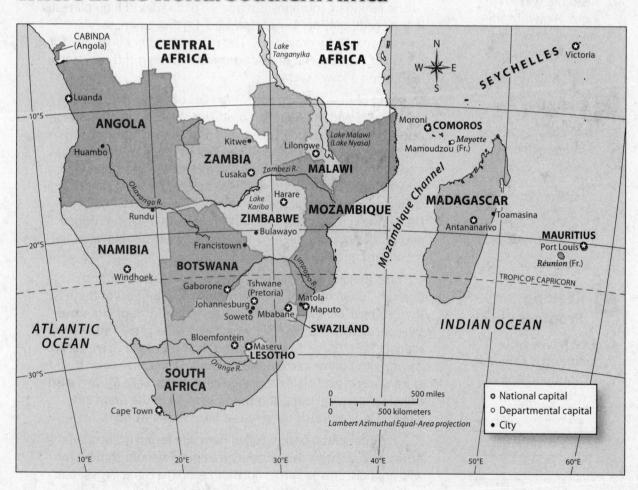

Southern Africa

Lesson 1: Physical Geography of Southern Africa, *continued*

? Identifying

1. Where do the Indian Ocean and the Atlantic Ocean meet?

_____

_____

_____

_____

✏ Marking the Text

2. Read the text on the right. Highlight the names of the three major river systems in Southern Africa.

A♭c Defining

3. How does a *reservoir* differ from a natural lake?

_____

_____

_____

_____

✓ Reading Progress Check

4. Which type of landform is common in Southern Africa?

_____

_____

_____

Landforms and Bodies of Water

Guiding Question *What are the dominant physical features of Southern Africa?*

Southern Africa includes the 10 southernmost countries on the African continent. It also includes four independent island countries and two French island territories off the east coast. It is bordered by the Indian Ocean on the east and the Atlantic Ocean on the west. Several of the countries, however, are **landlocked**, having no border with an ocean or sea. The two oceans meet at the Cape of Good Hope, on the southern tip of the continent.

Several of the countries are quite large. Angola and South Africa are each nearly the size of Western Europe. On the other end of the range, Madagascar is the world's fourth-largest island. The other island countries are tiny.

The region's geography is made up of a series of high plateaus. In the north, they are largely forested. Farther south, they are covered mainly by grasslands. The plateau's outer edges are called the Great Escarpment. An **escarpment** is a steep cliff between a higher and a lower surface.

A strip of desert called the Namib lies between the Great Escarpment and the Atlantic Coast. The Kalahari Desert is a vast, sand-covered plateau. It covers much of eastern Namibia and most of Botswana. South of the Kalahari Desert is a huge plateau that slopes from about 8,000 feet (2,438 m) in the east to 2,000 feet (610 m) in the west. At the southern tip of this plateau are several small, low mountain ranges called the Cape Ranges. The Drakensberg mountains in South Africa form the most rugged part of the escarpment. A narrow coastal plain lies between the mountains and the Indian Ocean.

Three major river systems—the Zambezi, Limpopo, and Orange—drain most of Southern Africa. The longest of these is the Zambezi. On the Zambia-Zimbabwe border, it plunges over the spectacular Victoria Falls, which are twice as tall and wide as Niagara Falls. These three rivers and their tributaries have carved canyons and gorges across the plateaus. Dams have been built to store water in artificial lakes called **reservoirs**. Lake Malawi (also called Lake Nyasa) is the southernmost lake of the Great Rift Valley. It is also one of the deepest lakes in the world.

A number of flat basins, called pans, are found in Southern Africa. They contain salt deposits that provide nourishment for wild animals. Etosha Pan in northern Namibia is the largest pan in Africa. It is at the center of Etosha National Park.

Southern Africa

Lesson 1: Physical Geography of Southern Africa, *continued*

Climate

Guiding Question *What is the climate of Southern Africa?*

The Tropic of Capricorn crosses the middle of Southern Africa. Northern Angola and northern Mozambique have a tropical wet-dry climate. Most rain falls in spring, summer, and fall—from October to May. The high elevation keeps temperatures cool. Along the coast of northern Mozambique, winds called monsoons bring heavy rains from the Indian Ocean during the summer.

Parts of Angola, Mozambique, Zimbabwe, Malawi, and Zambia have a humid subtropical climate with a shorter rainy season. Average temperatures here are slightly cooler. Nighttime frosts may occur in July in the high plateaus of Zambia and Malawi. In the lowlands, summer temperatures can exceed 100°F (38°C).

Most of the southern region has temperate climates that are not marked by temperature extremes. In much of South Africa, central Namibia, eastern Botswana, and southern Mozambique, summer days are warm, and winters are cool. There are frosts and sometimes freezing temperatures on the high plateaus. Annual rainfall varies, with most rain falling during the summer and very little during the rest of the year. Droughts are common and they can last for several years. Lesotho, Swaziland, and eastern South Africa, including the Indian Ocean coast, are much wetter.

Western South Africa, western Namibia, and much of Botswana are arid. Along the coast, the Namib gets little rain; in some years, no rain falls. Fog and dew provide some moisture, allowing small plants to survive. Inland areas of the Namib are hotter in summer and sometimes freeze in the winter. Because it is farther inland, the Kalahari has more extreme temperatures than the Namib. It also gets a little more precipitation than the Namib.

Southern African Climates	
Tropical wet-dry	northern Angola and northern Mozambique
Humid subtropical	southern Angola, western Mozambique, Malawi, Zambia, northeastern Zimbabwe
Temperate climates	South Africa, Lesotho, Swaziland, central Namibia, eastern Botswana, southern Mozambique
Arid climates	western South Africa, western Namibia, Botswana

Marking the Text

5. Read the text on the left. Underline the sentence that tells when most of the rain falls in areas with a tropical wet-dry climate.

? Explaining

6. How are plants able to grow in the Namib Desert, which receives little or no rain in a year?

_____

_____

_____

Comparing

7. Compare climates in Southern Africa's tropical, temperate, and arid regions.

_____

_____

_____

_____

✓ Reading Progress Check

8. Why are temperatures in Southern Africa's tropical countries generally not hot?

_____

_____

Southern Africa

Lesson 1: Physical Geography of Southern Africa, *continued*

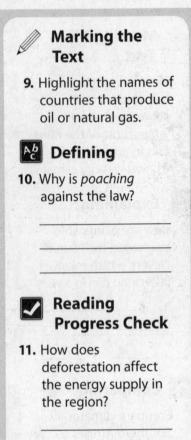

Marking the Text

9. Highlight the names of countries that produce oil or natural gas.

Defining

10. Why is *poaching* against the law?

_____

_____

_____

Reading Progress Check

11. How does deforestation affect the energy supply in the region?

_____

_____

_____

Natural Resources

Guiding Question *What natural resources are found in Southern Africa, and why are they important?*

Southern Africa is the continent's richest region in natural resources. The Republic of South Africa has some of the largest mineral reserves in the world. It is the world's largest producer of platinum, chromium, and gold, and one of the largest producers of diamonds. It also has important deposits of coal, iron ore, uranium, and copper, which have created a thriving mining industry.

South Africa, Zimbabwe, Botswana, and Mozambique mine coal and burn it to generate electricity. Mozambique and Angola have large deposits of natural gas. Angola is one of Africa's leading oil producers. Namibia also has oil and natural gas. The rivers and falls of Zimbabwe, Zambia, and Malawi are used to generate electricity. Deforestation, however, allows more sediment to enter rivers. This reduces water flow and the electricity that can be produced.

Namibia produces tin, zinc, copper, gold, silver, uranium, and diamonds. Rebels used Angola's diamonds to carry on a 20-year civil war. Diamonds used to pay for rebellions and other violent conflicts are called **blood diamonds**. Gold is a leading export for Zimbabwe. Zambia has some of the largest emerald deposits in the world. Malawi's most important natural resource is its soil. Its economy is based mainly on agriculture.

Southern Africa is known for its variety of animal life. Lions, zebras, giraffes, wildebeests, and other animals are found across the region. Nearly every country has created wildlife preserves to protect them. Tourists come from all over the world to see these animals. **Poaching**, or illegally killing game, is a problem.

Writing

Check for Understanding

1. **Informative/Explanatory** Compare and contrast the Namib and Kalahari deserts.

_____

_____

_____

_____

Southern Africa

Lesson 2: History of Southern Africa

ESSENTIAL QUESTION
How do new ideas change the way people live?

Terms to Know
apartheid the system of laws in South Africa aimed at separating the races
civil disobedience the act of disobeying certain laws as a means of protest
embargo a ban on trade with a particular country

When did it happen?

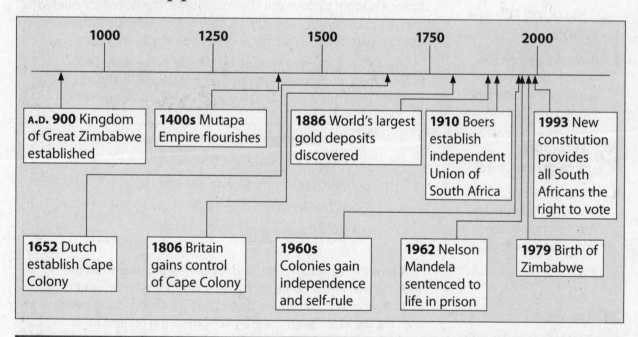

1000	1250	1500	1750	2000

A.D. 900 Kingdom of Great Zimbabwe established

1400s Mutapa Empire flourishes

1886 World's largest gold deposits discovered

1910 Boers establish independent Union of South Africa

1993 New constitution provides all South Africans the right to vote

1652 Dutch establish Cape Colony

1806 Britain gains control of Cape Colony

1960s Colonies gain independence and self-rule

1962 Nelson Mandela sentenced to life in prison

1979 Birth of Zimbabwe

Rise of Kingdoms

Guiding Question *What major events mark the early history of Southern Africa?*

South Africa's indigenous people have lived in the region for thousands of years. Some lived as hunter-gatherers. Others farmed and herded cattle. Trade among the groups flourished, as did trade with other countries. Ivory, gold, copper, and other goods moved from the interior to the east coast. There, goods were exchanged for tools, salt, and luxury items including beads, porcelain, and cloth from China, India, and Persia.

Marking the Text

1. Read the text on the left. Highlight the trade goods that originated in the region's interior.

Southern Africa

Lesson 2: History of Southern Africa, *continued*

Copyright © The McGraw-Hill Companies, Inc.

Sequencing

2. Describe the history of the Mutapa Empire.

_____

_____

_____

_____

_____

Marking the Text

3. Read the text on the right. Highlight the name of the ruler who formed the Zulu Empire.

Reading Progress Check

4. Which outsiders traded with Southern Africans before the Europeans arrived?

_____

_____

_____

Explaining

5. What was the purpose of Portugal's first settlements in Africa?

_____

_____

_____

_____

Around A.D. 900, the Shona people built a wealthy and powerful kingdom in what is now Zimbabwe and Mozambique. It was called Great Zimbabwe. By the 1300s, it had become a great commercial center, trading gold to Arabs at ports on the Indian Ocean. The city was abandoned in the 1400s. The growing population may have exhausted its water and food resources. Great Zimbabwe's ruins show the Shona's skill as builders.

In the late 1400s, the Shona conquered the region between the Zambezi and Limpopo rivers, from Zimbabwe to the coast of Mozambique. The Mutapa Empire traded gold for goods from China and India. In the 1500s, the Portuguese arrived and took over the coastal trade. In the late 1600s, Mutapa kings allied with the Rozwi kingdom to drive out the Portuguese. The Rozwi conquered the Mutapa's territory and ruled it until the early 1800s. It then became part of the Zulu Empire in what is now South Africa.

The Zulu leader Shaka united his people in the early 1800s to form the Zulu Empire. After he was killed in 1828, the empire survived until the British destroyed it in the Zulu War of 1879.

Kingdoms of Southern Africa

- Great Zimbabwe (A.D. 900–1400s)
- Mutapa Empire (late 1400s–late 1600s)
- Rozwi kingdom (late 1600s–early 1800s)
- Zulu Empire (early 1800s–1879)

A series of kingdoms rose and fell on Madagascar from the 1600s to the 1800s. Some of the early kingdoms were influenced by Arab and Muslim culture. In the early 1800s, one king allied with the British to keep the French from taking control of Madagascar. He conquered most of the island and formed the Kingdom of Madagascar. French troops invaded Madagascar in 1895 and made it a French possession.

European Colonies

Guiding Question *How did Southern Africa come under European control?*

Around 1500, Portugal and other countries started settlements along the African coast. At first, these were trading posts and supply stations at which ships could stop on their way to and from Asia. Eventually, Europeans grew interested in exploiting the region's natural resources.

netw✷rks

Lesson 2: History of Southern Africa, *continued*

In 1652, the Dutch founded Cape Colony on the Cape of Good Hope, at the southern tip of what is now South Africa. The Dutch became known as Boers, their word for farmers. They grew wheat and raised sheep and cattle. Enslaved people from India, Southeast Asia, and other parts of Africa did much of the work. The Africans fought the Dutch, but by the late 1700s they had been defeated. Some fled north, while others worked on the colonists' farms.

Wars in Europe gave Britain control of Cape Colony in the early 1800s. Thousands of British settlers soon arrived. The Boers resented British rule. Beginning in the 1830s, thousands migrated north of the Orange River in the Great Trek. In the 1860s, the Boers discovered diamonds. In 1886, they found the world's largest gold deposits. The British wanted control of these resources, which led to the Boer War in 1899. The Boers were defeated and again came under British control. In 1910 Britain allowed the Boer colonies to join Cape Colony. They formed the independent Union of South Africa. Britain maintained control of Lesotho and Swaziland.

Other European countries had been competing over the rest of Africa. In 1884 representatives of these countries met in Berlin, Germany, to divide the continent among themselves. The chart below shows how they divided Southern Africa.

European Colonies in Southern Africa	
Britain	Malawi, Zambia, Zimbabwe, Botswana, Mauritius, Seychelles
Portugal	Angola, Mozambique
Germany	Namibia (seized by South Africa during World War I)
France	Madagascar, Comoros

European control of Southern Africa continued for about 80 years. In the 1960s, the region's colonies began to gain independence and self-rule.

Independence and Equal Rights

Guiding Question *What challenges did Southern Africans face in regaining freedom and self-rule?*

In 1960 Madagascar became the first Southern African colony to gain independence. It was followed by most of the British colonies over the next few years. Portugal, however, refused to grant its colonies independence. In 1975, after the army overthrew the government, Angola and Mozambique became independent.

Marking the Text

6. Read the text on the left. Underline the sentence that explains who the Boers were.

Sequencing

7. Briefly describe Dutch and British attempts to control what is now South Africa.

_____

_____

_____

_____

Identifying

8. Name five present-day countries in Southern Africa that were once controlled by Britain.

_____

_____

_____

_____

Reading Progress Check

9. Which European country claimed the most territory in Southern Africa in the 1800s?

_____

networks

Southern Africa

Lesson 2: History of Southern Africa, *continued*

Marking the Text

10. Read the text on the right. Highlight the names of two Africans who became presidents of their countries after having been leaders of revolts against white rule.

✓ Reading Progress Check

11. Why do you think South Africa's government created the apartheid system?

_____

_____

_____

Rebel groups in each country fought for control until 1994 in Mozambique and 2002 in Angola.

White leaders in British Southern Rhodesia formed the country of Rhodesia. The country's African population carried on a guerilla war against the white rulers. In 1979, former rebel leader Robert Mugabe was elected president, and Rhodesia's name was changed to Zimbabwe.

The white minority government in South Africa stayed in power by limiting the black population's rights. The English controlled the government until the end of World War II. Then a strike by more than 60,000 black mine workers frightened the white population. In 1948 they elected an Afrikaner government that promised to take action. (Afrikaners are the descendants of the Boers.)

The new government created a system of **apartheid**—an Afrikaans word meaning "apartness." Laws forced black South Africans to live in areas called "homelands." The African National Congress (ANC) began a campaign of **civil disobedience**, disobeying certain laws as a means of protest. Armed conflict soon erupted. In 1962 ANC leader Nelson Mandela was arrested and sentenced to life in prison.

Other countries took note of events in South Africa. Some began to place **embargos**, or bans on trade, on South Africa. In 1989, the president was forced to resign. The following year, his successor began to repeal apartheid laws. Mandela was released from prison in 1991. In 1993 a new constitution gave all South Africans the right to vote, and in 1994 Mandela was elected president.

Writing

Check for Understanding

1. **Informative/Explanatory** Why did European countries meet in Germany to divide Africa?

_____

_____

2. **Argument** Why do you think that Rhodesia's new rulers named the country Zimbabwe?

_____

_____

_____

Southern Africa

Lesson 3: Life in Southern Africa

ESSENTIAL QUESTION
How does geography influence the way people live?

Terms to Know

utility services provided by the government such as electricity, water, and trash removal

thatch straw or other plant material used to cover roofs

periodic market an open-air trading market that springs up at crossroads or in larger towns

---

What Do You Know?

In the first column, answer the questions based on what you know before you study. After this lesson, complete the last column.

Now . . .		Later . . .
	What people live in Southern Africa?	
	What religions do Southern Africans follow?	
	How do the people of Southern Africa live?	
	What challenges do Southern Africans face?	

The People of the Region

Guiding Question *What people live in Southern Africa, and where do they live?*

The population of Southern Africa is overwhelmingly black. South Africa has the largest white minority. Whites make up 10 percent of the population there. In most other countries, whites and Asians make up less than 1 percent of the population.

Populations vary widely. Fewer than 2 million people live in Lesotho and Swaziland. Neighboring South Africa has the region's largest population, about 49 million. This is three times the population of Angola, which is about the same size. Most people in both countries live in cities. South Africa is the region's most industrialized nation. Malawi is the region's most densely populated country, with more than 250 people per square mile. It is also the region's most rural nation. Only 20 percent of its people live in cities. Malawi is the region's poorest country.

Activating Prior Knowledge

1. Why do you think South Africa has a higher population of whites than other countries in Southern Africa?

_____

_____

_____

_____

Lesson 3: Life in Southern Africa, *continued*

Marking the Text

2. Read the text on the right. Highlight the names of ethnic groups found in Southern Africa.

Determining Central Ideas

3. How did colonialism and contact with traders influence religious beliefs in Southern Africa?

_____

_____

_____

_____

_____

Reading Progress Check

4. What is the main religion practiced in Southern Africa?

_____

_____

Marking the Text

5. Read the text on the right. Underline the sentence that describes a Johannesburg "township."

Southern Africa is home to many ethnic and cultural groups who speak several different languages. The Shona make up more than 80 percent of the population of the country of Zimbabwe. South Africa's 9 million Zulu make them the country's largest ethnic group. More than 7 million Xhosa also live there. Some 4.5 million Tsonga people are spread among South Africa, Zimbabwe, and Mozambique. The San, a nomadic people, live mainly in Namibia, Botswana, and southeastern Angola. When they divided the region, Europeans often divided local peoples among several colonies.

Christian missionaries introduced Christianity to the region during the colonial era. Today, most of the people in almost every country are Christian. In Angola, however, nearly half the population continues to follow traditional African religions. In Zimbabwe and Swaziland, about half the population follows a blend of Christianity and traditional beliefs. There are large Muslim populations in the countries of Swaziland, Zambia, Malawi, and Mozambique. Contact with Arab traders long ago led to the introduction of Islam.

Portuguese is the official language in Angola and Mozambique. English is the official language in most of the former British colonies. However, most of the people speak indigenous languages. South Africa has 10 official languages besides English.

Life in Southern Africa

Guiding Question *How do the various people of Southern Africa live?*

Most of the people in Southern Africa live in the countryside. However, migration to cities is high because of job opportunities. Several cities in the region have more than 1 million people. Rapid growth of some cities has strained public **utilities**—services such as trash collection, sewage treatment, and water distribution. Outbreaks of cholera and other diseases can result from drinking polluted water.

Johannesburg, South Africa, has the most impressive downtown in all of Africa. White neighborhoods hold 20 percent of the city's population. Most black South Africans live in "townships" at the city's edge. These areas often have no electricity, clean water, or sewers. Most of the region's large cities have shantytowns.

Johannesburg is a mining, manufacturing, and financial center. It has attracted people from all over the world. Every black ethnic group in Southern Africa is present, as well. At least 12 languages are heard on the streets.

Southern Africa

Lesson 3: Life in Southern Africa, *continued*

Life in Johannesburg

• white neighborhoods outside the center city

• black "townships" at the city's edge

• is a mining, manufacturing, and financial center

• every black ethnic group in Southern Africa represented

• English and Afrikaner white population

• many European and Asian minority groups

In the countryside, traditional ways of life remain strong. Rural villages often consist of only 20 or 30 huts. They may be built of rocks, mud bricks, woven sticks and twigs packed with clay, and roofed in **thatch**—straw or other plant material. In most villages, all people are related by blood or marriage. Men often have more than one wife. They provide a hut for each wife and her children. Many families raise cattle for milk and as a symbol of wealth.

People in the countryside practice subsistence farming. Craftwork sometimes provides a source of cash. Men make wood and ivory carvings, and women make pottery to sell in cities or at **periodic markets**. These are open-air trading markets held regularly at crossroads or in larger towns.

In recent times, more men have been leaving their villages to work in cities or mines. The money they send home helps support their families. This trend has greatly changed village life. Many villages now consist largely of women, children, and older men. Women have increasingly taken on traditional male roles in herding, family and community leadership, and other activities.

Southern Africa Today

Guiding Question *What challenges and prospects do the countries of Southern Africa face?*

Southern Africa has great wealth, but also faces serious challenges. Life expectancy is low. Most people do not live beyond age 50 to 55. Lack of good rural health care is one reason. Many countries are trying to build or improve rural clinics. Diseases such as malaria, dysentery, cholera, and tuberculosis are widespread.

Malnutrition is a cause of death for many infants and young children. Southern Africa has some of the highest rates of infant death in the world. In Angola, Malawi, and Mozambique, about 100 to 120 of every 1,000 children die in infancy. Elsewhere, the figure is 40 to 60 per 1,000. (The infant death rate in the United States is 7 per 1,000.)

Defining

6. How do rural Southern Africans use *thatch*?

_____

_____

Describing

7. How has village life in Southern Africa changed in recent times?

_____

_____

_____

_____

_____

Reading Progress Check

8. Where in their countries do most Southern Africans live?

_____

_____

Comparing

9. Compare infant death rates in Southern Africa with those in the United States.

_____

_____

_____

_____

Southern Africa

Lesson 3: Life in Southern Africa, *continued*

🔲 **Analyzing**

10. How has HIV/AIDS affected the people of Southern Africa?

_____

_____

_____

☑ **Reading Progress Check**

11. Why is life expectancy in Southern Africa so low?

_____

_____

_____

A major cause of death in children and adults is HIV/AIDS. Southern Africa has the highest rate of HIV/AIDS infection in Africa. Swaziland, Botswana, Lesotho, and South Africa have the highest rates in the world. This sexually transmitted disease is also passed on by women to their children at birth.

Incidence of HIV/AIDS	
Swaziland, Botswana, Lesotho, South Africa	25 percent of adults
Rest of Southern Africa	11–14 percent of adults
United States	0.6 percent of adults

The high incidence of HIV/AIDS has disrupted the labor force by depriving countries of needed workers. It has also disrupted families through death, inability to work, or AIDS-related family issues. The disease has created millions of AIDS orphans. These are children whose parents have died from AIDS. The huge number of AIDS orphans is a major social problem.

Writing

Check for Understanding

1. **Informative/Explanatory** What are rural and city life like for Southern Africa's black population?

_____

_____

_____

_____

2. **Argument** What do you think is the biggest challenge the people of Southern Africa face?

_____

_____

_____

_____

networks

Australia and New Zealand

Lesson 1: Physical Geography

ESSENTIAL QUESTION
How does geography influence the way people live?

Terms to Know
Outback the most rural and isolated parts of Australia's Central Lowlands
monolith a single standing stone
Aboriginal people first humans to live in Australia
coral reef a giant community of marine animals called corals
hot spring pool of naturally occurring hot water
geyser spouts of hot water that shoot from the ground
drought long period of little or no rain
marsupial mammal that raises its young in a pouch on the mother's body
eucalyptus native Australian evergreen tree with stiff, pleasant-smelling leaves

Where in the World: Australia and New Zealand

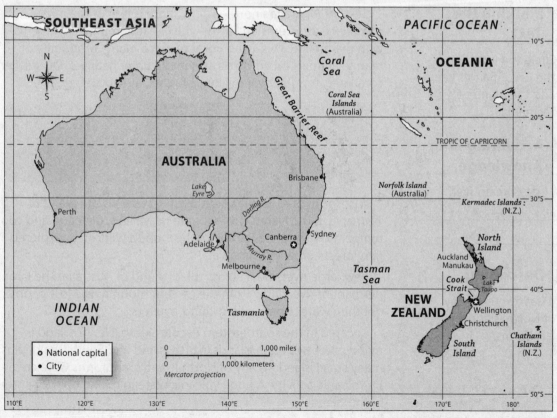

Australia and New Zealand

Lesson 1: Physical Geography, *continued*

 Describing

1. Why would it be hard for humans to live in the *Outback*?

_____

_____

_____

_____

Defining

2. Who are the *Aboriginal people*?

_____

_____

_____

Marking the Text

3. Circle the locations of the Great Barrier Reef and the island of Tasmania.

Activating Prior Knowledge

4. What is an aquifer?

_____

_____

Describing

5. What formed many of North Island's landforms?

_____

_____

The Land of Australia and New Zealand

Guiding Question *What physical features make Australia and New Zealand unique?*

Australia is the world's smallest continent. It is divided into three regions. The Western Plateau makes up the western half of Australia. It has rocky land with few water sources. Near the center are the interior highlands.

The Central Lowlands are a rough, dry region. Although rivers run through the area, it is mostly desert. The most rural and isolated parts of the Central Lowlands are called the **Outback**. In the Outback, strong winds cause harsh dust storms to blow across desert plains. Water is difficult to find.

In the Central Lowlands stands a massive, solid stone called a **monolith**. Many Australians call it Ayers Rock. The **Aboriginal people**, the first humans to live in Australia, call it *Uluru*. Uluru is sacred to many Aboriginal people. The Eastern Highlands, called the Great Dividing Range, run parallel to Australia's east coast. They are filled with mountains and valleys.

Just off Australia's northeastern shore is one of the most complex ecosystems on Earth—the Great Barrier Reef. It teems with marine life. This living **coral reef** is a giant community of marine animals called corals. It is made up of nearly 3,000 smaller coral reefs formed over millions of years, as shown below.

Corals lived, grew, and died. → Hard skeletons built up layer upon layer. → Large coral reefs formed.

Just south of Australia's southeastern coast is the island of Tasmania. It is covered with mountains, valleys, plateaus, and cool, temperate rain forests. Some of the world's wildest, unexplored rain forests grow in Tasmania.

Few large rivers flow across the dry land of Australia. The largest permanent rivers flow from the eastern mountains. Most people get their water from underground aquifers.

Located in the southeastern Pacific Ocean, New Zealand is made up of two main islands, North Island and South Island, plus many small islands. Many of North Island's landforms were created by volcanic activity. A huge volcanic plateau makes up the center of the island.

Australia and New Zealand

Lesson 1: Physical Geography, *continued*

Lake Taupo, New Zealand's largest lake, is in a massive crater formed by a volcanic eruption millions of years ago. The lake is surrounded by fertile plains and valleys. Northeast of Lake Taupo is an area famous for hot springs, mud pools, and geysers. **Hot springs** are pools of hot water that occur naturally. **Geysers** are huge spouts of hot water that shoot out of the ground. The water in hot springs and geysers is warmed by heat energy from deep within Earth.

New Zealand's South Island is famous for its Southern Alps. These towering mountains cover hundreds of miles along the western side of the island. The southern end of South Island is filled with a variety of landforms, including many fjords. These are deep glacier valleys filled with ocean water. New Zealand is located in the Ring of Fire. This is a region where active volcanoes and the frequent movement of tectonic plates often result in earthquakes.

Climates of the Region

Guiding Question *What types of climates and climate zones are found in Australia and New Zealand?*

The climate in Australia changes dramatically from one part of the country to another. The northern third of Australia has a warm, tropical climate. Winters are dry while summers are rainy and hot. Seasonal monsoons can bring heavy winds and rainfall.

The other two-thirds of Australia have a subtropical climate. Winters in Queensland, the Northern Territory, and Western Australia are warm and dry. Most rainfall occurs during spring and summer. Coastal areas tend to be sunny and dry with seasonal rains.

Much of central and western Australia has a desert climate with bands of semiarid steppes to the north, east, and south. Because Australia is the driest inhabited continent in the world, **drought** is a major problem. A drought is a long period of little or no rain. Droughts can threaten the survival of wildlife, livestock, and farm crops. In addition, low water reserves can lead to poor-quality drinking water for humans. Geographers can predict some droughts by monitoring climate changes caused by changes in global winds and ocean currents.

New Zealand's climate ranges from subantarctic in the south to subtropical in the north. Most areas on North Island and South Island have mild, temperate climates and both islands receive plenty of rain.

Defining

6. What is the difference between a *hot spring* and a *geyser*?

_____

_____

_____

_____

Reading Progress Check

7. In what ways are the lands of Australia and New Zealand alike, and in what ways are they different?

_____

_____

_____

_____

Reading Progress Check

8. What problems are caused by drought in Australia? Why are geographers monitoring climate changes there?

_____

_____

_____

_____

_____

networks

Australia and New Zealand

Lesson 1: Physical Geography, *continued*

 Marking the Text

9. In one color, highlight the type of animal that raises its young in a pouch. In another color, highlight the example of this type of animal that eats only eucalyptus leaves.

☑ **Reading Progress Check**

10. Explain why animals and plants native to Australia and New Zealand are different from living things in other parts of the world.

_____

_____

Plant and Animal Life

Guiding Question *What plants and animals are unique to Australia and New Zealand?*

Millions of years ago, the landmasses that became Australia and New Zealand separated from other lands. The animal and plant life developed in isolation from living things on other landmasses. Over millions of years, they adapted to live in their own unique environments. Many of New Zealand's animals, trees, ferns, and flowering plants are not found on other continents.

Australia's kangaroos, koalas, and bandicoots are **marsupials**. This is a type of mammal that raises its young in a pouch on the mother's body. Marsupials range in size from tiny kangaroo mice to 6-foot-tall kangaroos. Koalas have adapted to eat only one type of plant—the leaves of the **eucalyptus** tree. Eucalyptus are Australian evergreen trees with stiff, pleasant-smelling leaves. The leaves contain water, which can be difficult to find in drier regions. The table below lists some of New Zealand's unique animals.

Type of Animal	Example	Status
Lizard	geckos	common
Reptile	chevron skink	rare
Flightless birds	kiwi	common
	moa	extinct

Writing

Check for Understanding

1. **Narrative** Write a paragraph explaining what you would be interested in seeing on a trip to Australia or New Zealand. Be sure to include interesting landscapes, ecosystems, climates, and other features.

_____

_____

2. **Informative/Explanatory** List three features that make Australia unique and three that make New Zealand unique.

_____

_____

Australia and New Zealand

Lesson 2: History of the Region

ESSENTIAL QUESTION
Why does conflict develop?

Terms to Know

boomerang flat, bent, wooden weapon of the Australian Aborigines that is thrown to stun prey

dingo species of domestic dog first brought to Australia from Asia

tikanga traditional Maori customs and traditions passed down through generations

kapahaka traditional art form of the Maori people that combines music, dance, singing, and facial expressions

station cattle or sheep ranch in rural Australia

introduced species non-native animals brought from other places

dominion self-governing country within the British Empire

When did it happen?

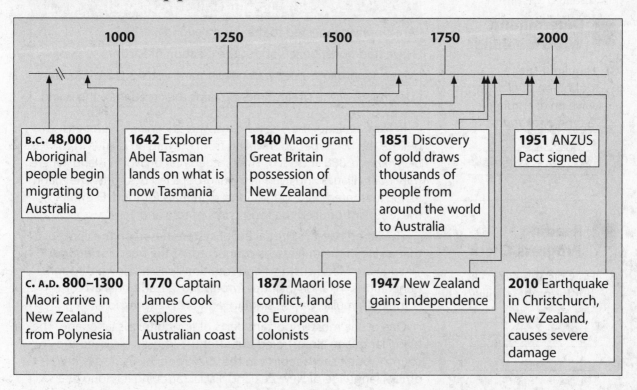

B.C. 48,000 Aboriginal people begin migrating to Australia

c. A.D. 800–1300 Maori arrive in New Zealand from Polynesia

1642 Explorer Abel Tasman lands on what is now Tasmania

1770 Captain James Cook explores Australian coast

1840 Maori grant Great Britain possession of New Zealand

1872 Maori lose conflict, land to European colonists

1851 Discovery of gold draws thousands of people from around the world to Australia

1947 New Zealand gains independence

1951 ANZUS Pact signed

2010 Earthquake in Christchurch, New Zealand, causes severe damage

Australia and New Zealand

Lesson 2: History of the Region, *continued*

 Summarizing

1. List two beliefs or traditions that most Aboriginal people share.

_____

_____

_____

_____

Marking the Text

2. Underline the text that explains how early Maori lived.

 **Determining Word Meaning**

3. Highlight the definition of *tikanga*. Give an example of a type of *tikanga*.

_____

_____

_____

Reading Progress Check

4. How did the early hunter-gatherers live?

_____

_____

_____

_____

First People

Guiding Question *How and when did the first humans settle in Australia and New Zealand?*

Fossil evidence shows that humans began migrating to Australia at least 50,000 years ago. These Aboriginal people were hunter-gatherers. They gathered fruit, roots, and other plant parts for food. They also hunted animals. They developed a flat, bent, wooden weapon called a **boomerang**. Hunters threw this to stun their prey.

Some hunter-gathers were nomadic and did not practice farming. They owned a type of animal called a **dingo**. Dingoes are a species of domestic dog brought to Australia about 4,000 years ago. (Today, wild dingoes roam free in the Outback and other parts of Australia.) Other Aboriginal people settled permanently in one location. Eventually they began farming.

Aboriginal people living in different parts of Australia developed their own languages, religions, traditions, and ways of life. Today they share some common beliefs and cultural traditions.

Aboriginal Beliefs and Traditions
Are closely connected to the natural world
Have traditional beliefs about the creation of Earth
Music, song, dance, poetry, storytelling, and visual arts are important
Dreaming, or the Dreamtime, explains the creation of the world

In New Zealand, the first humans were the Maori. Historians believe the Maori began traveling from Polynesian islands in canoes between A.D. 800 and 1300. They settled first on North Island, where they built villages and lived in tribal groups led by chiefs. The Maori fished, gathered plants, and hunted animals for food. Eventually, the Maori introduced the food crops of taro and yams.

The Maori have a spiritual belief system based on the concept that all life in the universe is connected. At the heart is **tikanga**. Tikanga are Maori customs and traditions passed down through generations. Tikanga includes **kapahaka**, a traditional art form combining music, dance, singing, and facial expressions.

One of the most important parts of any culture is language. The Maori language almost died out after the British colonized the country, but it saw a rebirth in the 1970s and 1980s. Today it is an official language of New Zealand. Facial tattooing has long been a unique part of Maori culture. Each tattoo was unique and showed the person's status, profession, and relationships. Other Maori art forms include weaving, painting, and wood carving.

Australia and New Zealand

Lesson 2: History of the Region, *continued*

Colonial Times

Guiding Question *What happened when Europeans came to Australia and New Zealand?*

During the 1600s, 1700s, and 1800s, Dutch, Spanish, French, and Portuguese explorers visited Oceania. Perhaps the most important was the British Captain James Cook. He explored the area in the 1760s and 1770s. After his voyages, the British government decided to send settlers to the wild lands of Australia. In 1788, a group of British ships brought convicted criminals from the British Isles. This was the first shipment of about 160,000 convicts sent to Australia during the next 80 years.

Beginning in the late 1700s, a settlement formed in Sydney, which grew into a busy center of trade and industry. In 1851, gold was discovered, and people from all over the world flocked to Australia. When coal, tin, and copper were discovered, workers came for mining jobs.

Over time, small towns grew into cities. Farmers planted crops and ranchers brought sheep and cattle from overseas. Vast ranches called **stations** covered millions of acres in the Outback and other areas.

As humans from other parts of the world moved to Australia, they brought along animals with which they were familiar. These are called **introduced species.** These are animals not native to an area but brought from other places. Ranchers brought dogs for guarding and herding sheep. Landowners brought European rabbits for hunting. Cane toads were brought to help get rid of a beetle that was destroying the sugarcane crop. Sometimes these new animals had devastating effects on Australia's environment.

Humans brought non-native animals to Australia. ➜ Introduced species crowded out and killed native species. ➜ Changes caused great damage to the environment.

British settlers often forced native Aboriginal people off their land. Also, European diseases reduced the Aboriginal population. Aboriginal survivors were forced to move to rugged lands that European settlers did not want.

In 1901, the six British colonies set up in Australia took action to unify as a federation. Together they formed the Commonwealth of Australia. This new country was a **dominion**, a largely self-governing country within the British Empire.

 Identifying

5. Who did the British bring to Australia in 1788?

_____

_____

_____

✎ **Marking the Text**

6. Highlight the four natural resources that brought people to Australia from all over the world.

 Describing

7. Describe what happened to the Aboriginal people following the arrival of British settlers.

_____

_____

_____

_____

_____

_____

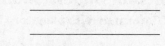 **Marking the Text**

8. Highlight the text that explains how Australia remained related to the British Empire.

Australia and New Zealand

Lesson 2: History of the Region, *continued*

9. Why did Australia, New Zealand, and the United States sign the ANZUS Pact?

_____

_____

_____

10. How and when did Australia and New Zealand become independent countries?

_____

_____

_____

_____

_____

In New Zealand, British colonists and British, American, and French traders and whalers built settlements on North Island. At first relations between the Maori and foreign settlers were peaceful.

In 1840, the British government convinced Maori leaders to sign a treaty giving legal ownership and control of New Zealand to Great Britain. As Europeans continued to arrive, the Maori saw more of their land taken by foreign settlers. Conflict between the Maori and British reached a head in 1872 when many Maori were killed. Most of their land was lost to the British.

As in Australia, businesses, industries, farms, and sheep ranches were built across New Zealand. Sheep ranching changed the land. In addition, introduced species such as rabbits, goats, and rodents destroyed natural habitats.

Throughout the 1800s, New Zealand residents pushed for independence from Great Britain. In 1907, the British government named New Zealand an independent dominion with a British-style parliamentary democracy. Even before independence, New Zealand was the first country in the world to legally recognize women's right to vote.

Australia and New Zealand were pulled into World War I and II because of their ties to Great Britain. After the wars, Australia became completely independent. In 1951, Australia signed a mutual security treaty with New Zealand and the United States called the ANZUS Pact. This treaty was meant to guarantee protection and cooperation among the three countries in case of military threats in the Pacific region.

Writing

Check for Understanding

1. Informative/Explanatory How are introduced species harmful to the land?

_____

_____

2. Informative/Explanatory In what ways were the colonization of Australia and New Zealand alike?

_____

_____

Australia and New Zealand

Lesson 3: Life in Australia and New Zealand

ESSENTIAL QUESTION
What makes a culture unique?

Terms to Know

bush large, undeveloped area where few people live

didgeridoo Aboriginal musical instrument that is a long wood or bamboo tube

action song performance combining body movement, music, and singing, often with lyrics celebrating Maori history and culture

geothermal energy naturally occurring heat energy produced by hot liquid rock in Earth's upper mantle

kiwifruit type of gooseberry fruit that is symbol of New Zealand

lawsuit a legal case brought before a court of law

What Do You Know?

In the first column, answer the questions based on what you know before you study. After this lesson, complete the last column.

Now . . .		Later . . .
	What religions are practiced in New Zealand?	
	What languages are spoken in New Zealand?	
	What kinds of natural resources do these countries have?	
	What environmental challenges do these countries face?	

Australia and New Zealand

Lesson 3: Life in Australia and New Zealand, *continued*

 Identifying

1. Which country has more people of European descent? Which has more native people?

_____

_____

✎ **Marking the Text**

2. Circle the percentage of people who live in cities and suburbs in both countries.

 Locating

3. How does rural life in Australia differ from rural life in New Zealand?

_____

_____

_____

_____

☑ **Reading Progress Check**

4. Based on information in the lesson, decide if the nicknames "Aussie" and "Kiwi" are offensive to Australians and New Zealanders, respectively.

_____

_____

_____

Life in the Region

Guiding Question *What is it like to live in Australia and New Zealand?*

Australia and New Zealand are multicultural lands. The table below shows the groups that make up the populations of the two countries.

Country	Group	Percentage of Population
Australia	European	92
	Asian	7
	Aboriginal and other groups	1
New Zealand	European	57
	Asian and Pacific Islander	12
	Maori	8
	Other groups	23

Christianity is the most common religion in the region. Buddhism, Islam, Hinduism, and native religions are also practiced. About a third of the people describe themselves as "nonreligious." English is the official language in Australia. New Zealand has three official languages: English, Maori, and New Zealand sign language.

Lifestyles in these countries are similar to modern American lifestyles. The residents drive cars, shop, watch television, and enjoy outdoor activities. The sports of rugby and football (soccer) are popular. Approximately 89 percent of Australia's people live in cities and suburbs. In New Zealand, about 87 percent of the people do so.

In rural areas of Australia, many individuals and families live alone on huge sheep and cattle stations. They are isolated and the work on the ranches is hard. The term **bush** means any large, undeveloped area where few people live, such as the Outback.

In New Zealand, rural areas are not as remote. Lush pasture lands can feed sheep and cattle on fewer acres than in the dry Australian Outback. New Zealand farms are closer together, and many people live near towns.

Australians are famous for their use of nicknames and slang. Common slang uses rhymes and substitutions. The Australian people call themselves "Aussies." The Australian nickname for New Zealanders is "Kiwis," which New Zealanders often call themselves.

Both Australia and New Zealand have well-respected universities that rank as some of the best schools in the world. Many students, especially in the Outback, use modern communication methods to receive and turn in their lessons.

Australia and New Zealand

Lesson 3: Life in Australia and New Zealand, *continued*

Australia and New Zealand have experienced revivals in Aboriginal and Maori cultures. An Aboriginal artifact still used today is the **didgeridoo**. This is a long wood or bamboo tube that creates an unusual vibrating sound when the player breathes into one end. In New Zealand, performers create **action songs** which combine body movements with music and singing. The lyrics often celebrate Maori history and culture.

Natural Resources and Economies

Guiding Question *What resources are important to the economies of Australia and New Zealand?*

Australia is rich in natural resources including coal, iron ore, gold, precious metals, oil, and natural gas. Fertile farmland is another valuable natural resource. Wool, food crops, and other agricultural products are raised across Australia. Timber and additional products come from trees. For example, eucalyptus trees are harvested for their wood, oil, resin, and leaves.

Australia's economy relies on exports of its natural resources. Coal, iron ore, and gold are its three leading exports. Meat, wool, wheat, and manufactured goods are also exported. Tourism is a vital industry. It brings revenue to local businesses and employs many Australians.

Because it is near the Ring of Fire, New Zealand uses **geothermal energy.** This is naturally occurring heat energy produced by extremely hot liquid rock in Earth's upper mantle. The heat energy is used to warm homes, provide hot water, and generate power.

Farmland is one of New Zealand's most valuable resources. Almost one-half of all land is used for farming and livestock grazing. Other valuable natural resources include coal, iron ore, natural gas, gold, timber, and limestone.

During the past two decades, New Zealand moved from a farm-based economy to an industrialized economy. Today, it produces wood and paper products, food products, machinery, and textiles. It also produces beef, lamb, fish, wool, wheat, flowering plants, vegetables, and fruit, such as the **kiwifruit**. This is a type of gooseberry that originated in East Asia and has become a symbol of New Zealand.

New Zealand is not as wealthy as Australia. However, its economy is 48th in the world in per capita gross domestic product (GDP). Like Australia, tourism is a major industry.

Defining

5. What is a *didgeridoo*?

_____

_____

_____

Marking the Text

6. Underline Australia's top three exports. Circle New Zealand's most valuable resource.

Defining

7. How is geothermal energy produced? How is it used in New Zealand?

_____

_____

_____

_____

_____

Reading Progress Check

8. Is the following statement a fact or an opinion? *Tourism is an important industry in Australia and New Zealand.*

_____

_____

Australia and New Zealand

Lesson 3: Life in Australia and New Zealand, *continued*

Drawing Conclusions

9. Why are more workers needed in Australia and New Zealand?

_____

_____

_____

_____

Reading Progress Check

10. In your own words, explain why survival of the Great Barrier Reef is threatened.

_____

_____

_____

_____

_____

_____

_____

Current Issues

Guiding Question *What challenges do the people of Australia and New Zealand face?*

The concern over Aboriginal people's rights has been an ongoing issue in Australia. Aboriginal activist groups have filed several major lawsuits over land rights and environmental issues. A **lawsuit** is a legal case that is brought before a court of law. The same type of human rights issues continue in New Zealand between New Zealanders of European descent and native Maori.

There are many environmental issues in Australia. These include drought, limited water supplies, bushfires, and threats to the Great Barrier Reef. All of these are affected by global warming.

Threats to the Great Barrier Reef
• Warming water temperature kills off organisms that need cooler temperatures.
• Corals are sensitive to climate change and water pollution.
• Parts of the Great Barrier Reef have already died.
• If global warming continues, the entire ecosystem may be lost.

In New Zealand, many geothermal hot springs and geysers have disappeared. Many of the remaining hot springs and geysers are located in protected areas. Geographers believe human activities such as drilling for hot water and building power stations have destroyed many natural wonders.

Another issue in Australia and New Zealand is the low birthrate combined with a gradually aging population. This creates a need for more workers to support the aging population. Immigrants are helping to fulfill that need.

Writing
Check for Understanding

1. **Argument** Think about issues facing Australians and New Zealanders today. Write a paragraph describing which issue you believe is the most important and why.

_____

_____

_____

netw⊙rks

Lesson 1: Physical Geography of Oceania

ESSENTIAL QUESTION
How does geography influence the way people live?

Terms to Know
continental island an island that was once connected to a larger continental landmass
archipelago a group of islands
high island an island with steep slopes rising from the shore and higher landforms
low island a relatively small, flat island with sandy beaches
lagoon a shallow pool in the center of an atoll
atoll a circular coral island surrounding a lagoon

Where in the World: Oceania

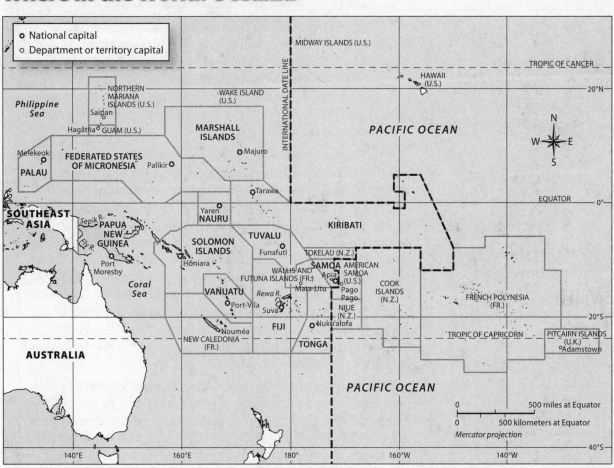

networks

Oceania

Lesson 1: Physical Geography of Oceania, *continued*

Landforms of Oceania

Guiding Question *How did thousands of islands appear across the Pacific Ocean?*

There are an estimated 10,000 islands in Oceania. Most of them are small, and many are uninhabited. Geographers divide the region into three sections, according to culture and physical location. These sections are shown in the chart below.

Oceania	
Micronesia	northwest section, east of the Philippines: Federated States of Micronesia, Palau, Guam, Marshall Islands
Melanesia	south of Micronesia, east of Australia: New Guinea, Solomon Islands, Vanuatu, Fiji, Tonga, Samoa
Polynesia	central Pacific Ocean: French Polynesia, Kiribati, Niue, the Hawaiian Islands, the Cook Islands

The islands range in size from New Guinea, which is 303,381 square miles (785,753 sq. km) to tiny islands covering less than 1 square mile (2.6 sq. km). Some, such as Fiji and Palau, are independent countries. Others, such as American Samoa and Guam, are overseas territories under the rule of other countries. Australia, England, France, New Zealand, and the United States have territories in Oceania.

New Guinea is the largest island in Oceania. It is a **continental island**, meaning it was once connected to a larger continental landmass. When sea levels were lower, it was a part of Australia. Now it is part of the Malay Archipelago. An **archipelago** is a group of islands. New Guinea is divided into two parts. The western part belongs to Indonesia. The eastern part is an independent country called Papua New Guinea. Only Papua New Guinea is considered part of Oceania. In this discussion, we will refer to the entire island as Papua New Guinea. The island has a great many landforms. Rugged mountain peaks and glaciers dominate the center of the island. There are low mountains and fertile river valleys in the north, and a northern coastal plain. In the south, swampy lowlands lead to the Owen Stanley Range.

Oceania includes several types of smaller islands. **High islands** have steep slopes rising from the shore, higher landforms, and diverse plant and animal life. They are usually the greenest of the small islands. Fertile soil and humid and rainy climates allow dense rain forests to grow. Many have freshwater streams, rivers, and waterfalls. Tahiti and the Hawaiian islands are high islands.

Marking the Text

1. Read the text on the right. Highlight the names of countries that have territories in Oceania.

Defining

2. Why is New Guinea called a *continental island*?

_____

_____

_____

Describing

3. What is the difference between New Guinea and Papua New Guinea?

_____

_____

_____

_____

Comparing and Contrasting

4. In what ways are low islands and high islands alike and different?

_____

_____

_____

_____

_____

Oceania

Lesson 1: Physical Geography of Oceania, *continued*

Most high islands of Oceania were formed by underwater volcanoes. As lava from erupting volcanoes flowed into the ocean waters, it cooled and formed large mounds of volcanic rock. Eventually the volcanic rock built up above the water's surface, forming high islands.

Low islands are small, flatter islands with sandy beaches. They tend to have fewer forests and less diverse plant and animal life. Most of the islands of Micronesia and Polynesia are low islands. Some of them are so low that they just break the water's surface. The island country of Tuvalu is so low that it is being eroded by ocean waters. If sea levels continue to rise, Tuvalu and Oceania's other lowest islands will disappear under the water.

Low islands were formed from coral reefs circling underwater volcanoes. These reefs eventually grew above the water's surface. In the center of these reefs were shallow pools called **lagoons**. As the reefs aged and crumbled, ocean waves deposited sediment on the coral remains, forming an **atoll**. An atoll is a coral island made up of a reef island surrounding a lagoon.

Climates of Oceania

Guiding Question *What factors affect climate in Oceania?*

Nearly all of Oceania is located within the Tropics. Thus, most of the islands have warm, humid, tropical climates. Those outside the Tropics have mixed tropical and subtropical climates. Some islands experience local climate variations caused by elevation, winds, and ocean currents.

Papua New Guinea has two climate zones: tropical and highland. The highland regions are much cooler. The climate is wet. February has the most rain, and July has the least. Monsoon rains are common. Much of the island is covered in trees and other forest plants. There are, however, occasional droughts caused by the El Niño effect.

Temperatures on the smaller islands of Oceania are generally warm throughout the year. They receive large amounts of rain. Seasonal rainfall patterns determine an island's wet season. That is the time of the year when the heaviest rains fall. Islands north of the Equator, such as the Marshall Islands and Palau, have a wet season from May to November. Islands south of the equator, such as Samoa and Tonga, have a wet season from December to April. These rainfall patterns occur under normal conditions. Heavy rains can also be caused by storms such as typhoons. Typhoons cause high winds and powerful waves that can topple trees and houses and erode shores. The only small island affected by monsoon winds is Yap.

☑ Reading Progress Check

5. What are the names of Oceania's sections?

_____

_____

_____

✎ Marking the Text

6. Read the text on the left. Underline the sentence that tells how El Niño affects one of the countries in Oceania.

⚙ Analyzing

7. What factors affect the climates of Oceania?

_____

_____

_____

_____

_____

☑ Reading Progress Check

8. The Northern Mariana Islands experience a wet season from May to November. Based on this information, are the Northern Mariana Islands located north or south of the Equator?

_____

Oceania

Lesson 1: Physical Geography of Oceania, *continued*

<table>
<tr><td>

⚙ Analyzing

9. Why are there few natural resources on low islands?

_____

_____

_____

_____

✎ Marking the Text

10. Read the text on the right. Underline the sentence that describes products from Oceania's high islands.

☑ Reading Progress Check

11. What are two types of renewable resources in Oceania?

_____

_____

_____

_____

</td><td>

Resources of Oceania

Guiding Question *What natural resources do the islands of Oceania possess?*

Since most of Oceania's islands are small, the amount of natural resources is limited. In spite of this, they have some valuable resources. Those that are sold or traded are essential to the islands' economies.

The nation of Papua New Guinea has many natural resources. It has large natural gas reserves for its size. They have the potential to benefit the nation's economy.

Natural Resources of Papua New Guinea		
gold	timber	petroleum
copper	fish	natural gas

Oceania's smaller islands have few natural resources that can be traded on the international market. Limited land area is one reason. Low islands, based on coral, do not have rock foundations. It is in deep layers of rock that deposits of metal ores are found.

New Caledonia is developing wind power. That form of power is also spreading to other islands. Kiribati and the Solomon Islands have begun using solar-powered lighting in their homes. Wind and solar energy are renewable, nonpolluting resources. Because of the sunny climates and ocean winds, these resources are plentiful throughout Oceania.

Some of the high islands have trees that are used for timber, rubber, and other products. Soil quality varies from island to island. Some islands have rich volcanic soil for growing farm crops. Other islands have poor-quality soil, which makes farming difficult. Fish and other seafood are important resources. Most islands use fish only for their own food, not for export.

</td></tr>
</table>

Writing

Check for Understanding

1. **Informative/Explanatory** How did thousands of islands appear across the Pacific Ocean?

_____

_____

Oceania

Lesson 2: History and People of Oceania

> **ESSENTIAL QUESTION**
> *What makes a culture unique?*

> **Terms to Know**
> **wayfinding** ancient method of navigation that relies on observation of the natural world, such as the sun, stars, or ocean currents
> **trust territory** an area temporarily placed under control of another country by the United Nations
> **possession** an area or a region that is controlled by a foreign government
> **pidgin language** a language formed by combining parts of several different languages
> *fale* a traditional Samoan home made of wood poles with a thatched roof that has no walls

When did it happen?

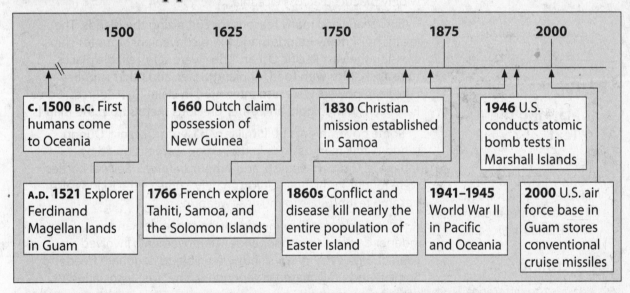

| 1500 | 1625 | 1750 | 1875 | 2000 |

c. 1500 B.C. First humans come to Oceania

1660 Dutch claim possession of New Guinea

1830 Christian mission established in Samoa

1946 U.S. conducts atomic bomb tests in Marshall Islands

A.D. 1521 Explorer Ferdinand Magellan lands in Guam

1766 French explore Tahiti, Samoa, and the Solomon Islands

1860s Conflict and disease kill nearly the entire population of Easter Island

1941–1945 World War II in Pacific and Oceania

2000 U.S. air force base in Guam stores conventional cruise missiles

History of Oceania

Guiding Question *How were the islands of Oceania populated?*

The first humans began settling in Oceania around 1500 B.C. Historians believe the early settlers came from Southeast Asia. Over many centuries, people from areas such as the Philippines and Indonesia sailed from their homelands and settled the islands across Oceania. Settlers from other islands then migrated farther into Oceania. Their only way these settlers had to navigate their boats across the water was by **wayfinding**.

> **?** **Explaining**
>
> 1. For approximately how long have humans been living on the islands of Oceania?
>
> _____
>
> _____

networks

Oceania

Lesson 2: History and People of Oceania, *continued*

Marking the Text

2. Read the text on the right. Underline the sentence that explains where Oceania's original settlers came from.

? **Describing**

3. How did World War II affect the islands of Oceania?

_____

_____

_____

_____

_____

Marking the Text

4. Read the text on the right. Circle the names of foreign governments that have agreements with islands of Oceania.

A^b_c **Defining**

5. What is the difference between a *trust territory* and a *possession*?

_____

_____

_____

_____

_____

Wayfinding

An ancient method of navigation that:

• relies on careful observation of the natural world.

• uses the sun, the stars, and the movement of ocean currents and swells to chart courses.

• was practiced long before the invention of navigation instruments, such as compasses and sextants.

European explorers began sailing through Oceania in the 1500s. They began to colonize the islands in the 1600s. Violent conflict often broke out between colonists and native peoples. During the 1800s and 1900s, Christian missionaries came to the islands. Native people did not always welcome them. However, many local populations converted to Christianity. Christian faiths are still widely practiced across Oceania.

Europeans had many reasons for colonizing the islands. The locations of many islands made them convenient stops for ships crossing the vast Pacific Ocean. They were safe, reliable places to restock ships with food, drinking water, and other supplies. European powers were also interested in claiming resources. Some Europeans mined gold and other precious metals from the islands.

Some governments built military bases in Oceania. During World War II, many islands in the region were occupied by Japanese, German, English, and American forces. Several battles were fought in the area. Unfortunately, some of the islands also became testing sites for nuclear weapons.

After World War II, many islands began to demand independence. Some independence movements involved conflict. However, many islands were able to negotiate freedom. Some islands negotiated independence by "free association" with foreign powers.

The island of Palau, for example, is an independent republic but has a voluntary free association with the United States. Palau allows the United States to keep military facilities on one of its islands. In return, the United States provides millions of dollars of aid to Palau each year.

Other islands have agreements of various kinds with foreign governments, including Australia, New Zealand, Great Britain, France, and the United States. They generally involve use of land or other resources in exchange for military protection and economic aid.

Oceania

Lesson 2: History and People of Oceania, *continued*

A **trust territory** is one that has been placed under the authority of another country by the Trusteeship Council of the United Nations. The Marshall Islands were a trust territory until they gained independence in 1986. A **possession** is a territory occupied or controlled by a foreign government. French Polynesia is a possession because it is an overseas territory of France. Trust territories and possessions do not govern themselves but are run by foreign governments.

The People of Oceania

Guiding Question *What is life like in Oceania?*

Oceania has one of the world's most diverse populations. Many islands are home to a wide range of ethnic groups. The most amazing diversity is found in Papua New Guinea. It has a total population of more than 6 million in an area about the size of California. The nation's population is made up of people from many ethnic and native tribal groups, as shown in the graph below.

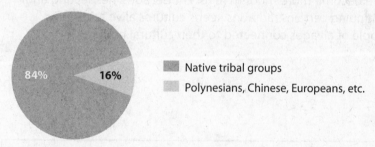

84% 16%

■ Native tribal groups
■ Polynesians, Chinese, Europeans, etc.

The native population includes people from hundreds of different tribal groups. The different groups have their own lifestyles, cultural traditions, beliefs, and languages. As many as 860 different languages are spoken in Papua New Guinea. That is 10 percent of all the languages known to exist.

In spite of their many languages, the people of Papua New Guinea are still able to speak to one another. Many speak a pidgin language as well as their own language. A **pidgin language** is a language used for communication between people who speak different languages.

Such wide diversity has both positive and negative effects. Papua New Guinea's many different ethnic groups create a rich and varied culture. At the same time, there are problems such as ethnic discrimination and violent conflicts among tribal groups. Only about 12 percent of the population lives in urban areas. Thousands of distinct tribal groups live in small villages in the nation's many remote locations.

Reading Progress Check

6. Why did the first Europeans come to Oceania?

_____

_____

_____

Determining Central Ideas

7. Describe the diversity found in Papua New Guinea's population.

_____

_____

_____

Marking the Text

8. Highlight the number of languages spoken in Papua New Guinea. What percentage of the world's languages is that?

_____

Making Connections

9. Read the text on the right. Underline the sentence that describes the positive effects of diversity in Papua New Guinea.

Oceania

Lesson 2: History and People of Oceania, *continued*

? **Describing**

10. How have modern influences changed life in Oceania?

_____

_____

_____

_____

✓ **Reading Progress Check**

11. Why might using a pidgin language be useful to the population?

_____

_____

_____

Life in the villages of Oceania is based on tradition. One tradition is an important event in the lives of young people called the coming-of-age ceremony. Ceremonies and celebrations include feasts and dancing. Like their ancestors, Polynesian people practice artistic wood carving. Many Polynesian and Micronesian cultures practice tattooing. Micronesian people use storytelling to retell history and keep track of family heritage. In Samoa, traditional homes called **fales** are common. These are open structures made of wood poles with thatched roofs made of palm leaves. They are used mainly for shade and shelter from frequent rainfall.

Today, Oceania's cultures are a blend of traditional and modern practices, beliefs, and lifestyles. Christianity is widely practiced, along with elements of traditional religions. Many people wear Western-style clothing and hairstyles. Cell phones and laptop computers are common. Traditional celebrations are practiced throughout Oceania. Some of these, such as the Hawaiian luau, have become more modern in recent decades. Respecting and continuing certain traditions keeps cultures alive and keeps people of all ages connected to their cultural heritage.

Writing

Check for Understanding

1. Informative/Explanatory How were the islands of Oceania populated?

_____

_____

_____

2. Informative/Explanatory List three traditions still practiced in Oceania.

_____

_____

_____

netw⚙rks

Oceania

Lesson 3: Life in Oceania

ESSENTIAL QUESTION
Why do people make economic choices?

Terms to Know

cash crop a farm product grown to sell for profit
resort tourist areas often located in beautiful areas
remittance foreign-earned wages that are sent back to a home country
MIRAB economy a lesser-developed economy that depends on aid from foreign countries and remittances from former residents working elsewhere

---

What Do You Know?

In the first column, answer the questions based on what you know before you study. After this lesson, complete the last column.

Now . . .		Later . . .
	How do the people of Oceania earn their livings?	
	What obstacles to economic development exist in Oceania?	
	What challenges do the people of Oceania face?	

The Economies of Oceania

Guiding Question *How do the people of Oceania earn their livings?*

Papua New Guinea has the most valuable natural resources in Oceania, after Australia and New Zealand. Gold and copper are the country's most profitable resources. They account for about 60 percent of the country's export income. Other major exports are silver, timber, and cash crops. **Cash crops** are crops grown or gathered to sell for profit.

Marking the Text

1. Read the text on the left. Highlight the major exports of Papua New Guinea.

Oceania

Lesson 3: Life in Oceania, *continued*

❓ Describing

2. How do the majority of people in Papua New Guinea make their living?

_____

_____

❓ Explaining

3. Why do so many of the nations of Oceania have to import more than they export?

_____

_____

_____

_____

✏ Marking the Text

4. Read the text on the right. Highlight the names of countries that pay for fishing rights in Oceania.

🔤 Explaining

5. How important is tourism to the economies of many small countries in Oceania?

_____

_____

_____

_____

Many people have jobs in various industries. However, the vast majority of Papua New Guinea's people live by subsistence farming. Although unemployment is low, most people earn low incomes. The government plans to increase exports of minerals and petroleum to strengthen the country's economy.

The Economy of Papua New Guinea		
Cash Crops	**Subsistence Crops**	**Industries**
Rubber	Yams	Mining
Tea	Sweet Potatoes	Farming
Coffee	Taro	Timber Processing
Cacao	Bananas	Palm Oil Refining
Coconuts	Pigs	Petroleum Refining
	Chickens	Tourism

Smaller independent island countries throughout Oceania face many obstacles to economic development. They have limited land, poor soil, and large populations. They must import much of their food, fuel, finished goods, and raw materials. Most import far more than they export. Many import five or six times as much as they export. In Tuvalu, for example, imports exceed exports by 200 to 1.

People on smaller islands raise what food they can by subsistence farming. Some islands raise limited cash crops, such as fruits, vegetables, sugar, nuts, coffee, tea, cocoa, and palm and coconut oils. Farms and plantations employ some residents, but the work is exhausting and pays little.

Most fishing operations in Oceania are small. Fishers catch only enough to feed their families or to sell to local markets. Most island countries do not have equipment or processing facilities to operate large fishing industries. Some of them sell fishing rights to other countries. Japan, Taiwan, South Korea, and the United States pay for access to Oceania's marine resources.

On islands that have minerals and other marketable resources, many people work for industries such as mining, fishing, clothing, and farming. Some people make a living by using local materials to create art and to craft tools and artifacts.

Tourism is important to the economies of many small, independent countries in Oceania. Tourists come from all over the world to enjoy the sunshine, warm waters, and beautiful views. **Resorts** provide comfortable lodging, food, recreation, and entertainment. Without the revenue from tourism, many small islands would be at risk of economic collapse.

Lesson 3: Life in Oceania, *continued*

Many people on islands with less-developed economies depend on money from family members living overseas. Foreign-earned wages, called **remittances**, are vital sources of income for families. Many of the islands have what are called **MIRAB economies**.

MI	**MI**gration
R	**R**emittances
A	**A**id
B	**B**ureaucracy

Islands with MIRAB economies depend on outside aid and remittances. One example is the Polynesian nation of Tonga. Tonga raises a few export crops, such as squash, vanilla beans, and root vegetables. The tourism industry also brings some revenue. However, 75 percent of all Tongan families receive remittances.

Tonga and many other small island countries also depend on aid from foreign governments. Australia, Great Britain, Japan, New Zealand, and the United States have created trust funds for many islands. Foreign aid is used for food, schools, water and sanitation, and other basic needs. It also funds economic development. U.S. territories such as Guam and Wake Island are home to U.S. military bases. Local people are employed in industries that serve the needs of military personnel.

Issues Facing the Region

Guiding Question *What challenges do the people of Oceania face?*

The future is uncertain for the islands of Oceania. Migration is a major issue. So many young people have left Oceania to seek work overseas that the populations of some islands have changed dramatically. Half of all Tongan people live abroad. A large percentage of the people remaining in Tonga are children or older people unable to work. More and more Tongans are born in other countries. Before long, the majority of Tongan people will be born and raised overseas, weakening cultural ties to the homeland.

Another major issue is the need for economic development. On many islands, economic development is slow because of the lack of resources. As people migrate to cities to try to find work, fewer people are growing their own food. Many countries of Oceania import more goods and services than they export. When a country imports more than it exports, the entire economy is affected. Countries like to export more than they import because this creates jobs and demand for goods and services.

Marking the Text

6. Read the text on the right. Highlight the phrase that explains what remittances are.

Reading Progress Check

7. In your own words, briefly explain MIRAB economies.

_____

_____

_____

_____

Analyzing

8. The birthrate in Tonga has increased, but the island's population has decreased. Explain.

_____

_____

_____

_____

Explaining

9. Why do countries prefer to export more than they import?

_____

_____

_____

_____

Oceania

Lesson 3: Life in Oceania, *continued*

Drawing Conclusions

10. What do you think is the most serious issue or challenge in Oceania today? Use information from the lesson in your answer.

_____

_____

_____

✓ Reading Progress Check

11. Why is climate change a serious concern in Oceania?

_____

_____

Most of the nations of Oceania are free of conflict, but there is some unrest. Crime and human rights abuses are problems on some islands. In Fiji, tensions between natives and immigrants from India led to conflicts. The Solomon Islands have also seen conflict. The native people of Guadalcanal and people from the neighboring island of Malaita disagreed on land rights, but peace was restored with the help of the United Nations.

Oceania faces serious environmental issues. These include climate change, deforestation, pollution, natural disasters, and declining fish populations. Climate change is one of the most serious issues in the region. With continued global warming, sea levels are rising. If this continues, many of the low islands could be completely covered by water. Some of the lowest islands are already experiencing surface flooding from rising oceans and eroding beaches.

Natural disasters such as earthquakes, tsunamis, typhoons, and resulting floods have the potential to destroy homes and claim lives. They also damage farmland and natural habitats.

Commercial fishing companies have harvested so many fish from some areas that almost no fish remain for local people to catch and eat. When huge numbers of fish are caught at once, the remaining fish are unable to reproduce fast enough to restore populations. In time, entire fish populations will be gone. Ocean pollution also threatens the survival of fish and other marine animals.

Writing
Check for Understanding

1. Informative/Explanatory List three ways the people of Oceania earn a living.

_____

_____

_____

2. Argument Do remittances benefit or harm cultures and economies? Explain your answer.

_____

_____

_____

networks

Antarctica

Lesson 1: The Physical Geography of Antarctica

ESSENTIAL QUESTION
How does physical geography influence the way people live?

Terms to Know
ice sheet thick layer of ice and compressed snow that forms a solid crust over an area of land
ice shelf thick layer of ice that is attached to a coastline and floats on the ocean
calving process of ice breaking free from an ice shelf or glacier
iceberg large chunk of floating ice that broke off from an ice shelf or glacier
katabatic wind powerful, cold wind that blows from the interior of Antarctica
lichen organisms made up of algae and fungi that grow on rocky surfaces
krill tiny shrimp-like sea creatures that provide food to whales and many other sea creatures
plankton tiny organisms—algae or single-cell bacteria—that float near the water's surface

Where in the World: Antarctica

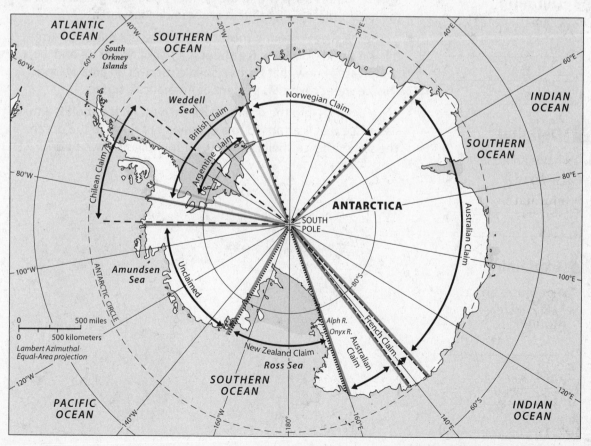

Antarctica

Lesson 1: The Physical Geography of Antarctica, *continued*

Landforms and Waters of Antarctica

Guiding Question *Is Antarctica the last unknown land region on Earth?*

The continent of Antarctica is nicknamed the "last continent" and "the bottom of the world." It lies more than 2,500 miles (4,023 km) south of the Tropic of Capricorn. It is the world's fifth-largest continent and is surrounded on all sides by the Southern Ocean. The South Pole is located near the center of the continent.

Antarctica is the world's highest continent. It has an average elevation of 7,000 feet (2,134 m) above sea level. However, most of the surface we can see is packed ice. Some of Antarctica's mountains rise above the ice, but most of the land is much lower. In fact, the weight of this ice has forced some of Antarctica's surface land far below sea level.

The highest point in Antarctica is Vinson Massif, which reaches 16,066 feet (4,897 m) in height. The Transantarctic Mountains divide the continent into East Antarctica and West Antarctica. The lowest point is deep within the Bentley Subglacial Trench. This is the lowest place on Earth that is not under seawater.

The ice sheet covering most of Antarctica is 2 miles (3.2 km) thick in some places. An **ice sheet** is a thick layer of ice and compressed snow that forms a solid crust over an area of land. This ice sheet covers 98 percent of Antarctica's surface.

Ice extends into the ocean on ice shelves. An **ice shelf** is a thick slab of ice that is attached to a coastline but floats in the ocean. The size of the ice shelf changes year round, as shown below.

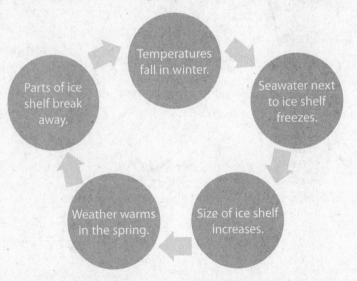

- Temperatures fall in winter.
- Seawater next to ice shelf freezes.
- Size of ice shelf increases.
- Weather warms in the spring.
- Parts of ice shelf break away.

? Explaining

1. Why is some of Antarctica's land below sea level?

_____

_____

_____

✎ Marking the Text

2. Read the text and circle the highest and lowest points in Antarctica.

A͚ᵇ꜀ Defining

3. What is an *ice sheet*?

_____

_____

_____

A͚ᵇ꜀ Defining

4. What is an *ice shelf*?

_____

_____

_____

? Explaining

5. What causes an ice shelf to increase in size?

_____

_____

_____

Antarctica

Lesson 1: The Physical Geography of Antarctica, *continued*

The process of ice breaking free from an ice shelf or glacier is called **calving**. The large chunk of ice that breaks off is called an **iceberg**. Icebergs are made of freshwater, not salt water. To be called an iceberg, the chunk of ice must be at least 16 feet (5 m) across. As much as 90 percent of an iceberg remains underwater. This has led to accidents when ships have come close to what appeared to be small chunks of ice. The ships then crashed into the enormous, unseen parts of the icebergs.

Antarctica is surrounded by several seas: the Weddell Sea, the Scotia Sea, the Amundsen Sea, the Ross Sea, and the Davis Sea. These were named by and for explorers and scientists who mapped and studied the region.

Climate and Resources

Guiding Question *Does any life exist on such a forbidding continent?*

Antarctica has an intensely cold climate. Because of its high latitude, it never receives direct rays from the sun. Thus, very little heat energy reaches the land. In addition, for three months of the year, most of Antarctica receives no sunlight at all.

The high elevation also affects the climate. Strong, fast winds blow colder air down from high interior lands toward the coasts. These powerful gusts are called **katabatic winds**. They are driven by Earth's gravity. In addition, Antarctica has an extremely dry climate. It only receives 2 to 4 inches (5 to 10 cm) of precipitation per year, making it the world's driest continent.

Antarctica plays an important role in maintaining the global climate balance. Its ice covering affects the temperature of Earth, as shown below.

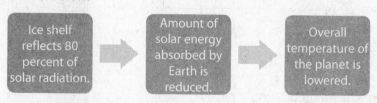

Antarctica's harsh climate limits the types of plants and animals that can live on the surface. Only 1 percent of Antarctica's land is suitable for plant life. Plants found here include flowering plants, mosses, algae, and **lichens**. Lichens are organisms that usually grow on solid, rocky surfaces and are made up of algae and fungi. Most plants grow on the Antarctic Peninsula and surrounding islands. These are the warmest and wettest places in the region.

Defining

6. What is the name of the process that forms icebergs?

_____

Marking the Text

7. Highlight the text that explains why icebergs cause accidents.

Reading Progress Check

8. Does the continent of Antarctica increase in size during the winter?

_____
_____
_____
_____

Explaining

9. Why are there so few plants in Antarctica?

_____
_____
_____
_____

Defining

10. Name two types of organisms that make up *lichens*.

_____
_____

Antarctica

Lesson 1: The Physical Geography of Antarctica, *continued*

Marking the Text

11. In one color, highlight the life-form that sea mammals and seabirds eat. In another color, highlight what is eaten by this life-form.

Reading Progress Check

12. In your own words, explain why Antarctica has such a cold, dry climate.

_____

_____

_____

_____

Most land animals are tiny insects and spiderlike mites. Most other land animals are only visitors, such as Weddell seals and many species of shorebirds, such as albatross, cormorants, and gulls. Only the emperor penguin stays year round. These giant penguins have adapted to the cold temperatures and cannot survive outside of Antarctica's frozen climate.

Although the land is home to very few living things, the seas are filled with life. Many kinds of seals, dolphins, fish, and other marine animals live in the waters around Antarctica. Whales come to the Southern Ocean to feed during the summer. Sea mammals and seabirds eat **krill**, tiny crustaceans similar to shrimp. Krill thrive in the Southern Ocean and feed on a smaller life-form called **plankton**. These are tiny organisms such as single-celled bacteria or algae that float near the water's surface. Plankton, krill, fish, sea mammals, seabirds, and land animals are all part of Antarctica's ecosystem. They are an essential part of the food chain that supports life on the continent.

It is difficult or impossible to explore Antarctica for natural resources. The harsh climate makes it dangerous for human workers, and the land is buried by the thick ice sheet. As a result, mineral and energy resources that might exist have not been exploited. However, commercial fishing operations harvest fish from the water.

Writing

Check for Understanding

1. Informative/Explanatory How does Antarctica help lower Earth's temperature?

_____

_____

_____

2. Informative/Explanatory Explain why Antarctica's mineral and energy resources are not commercially mined or harvested.

_____

_____

_____

Antarctica

Lesson 2: Life in Antarctica

ESSENTIAL QUESTION
How do people adapt to their environment?

Terms to Know

treaty agreement or contract between two or more nations, governments, or political groups

research station base where scientists live and work

ozone gas in the atmosphere that absorbs harmful ultraviolet radiation from the sun

remote sensing use of scientific instruments placed onboard weather balloons, airplanes, and satellites to study the planet

When did it happen?

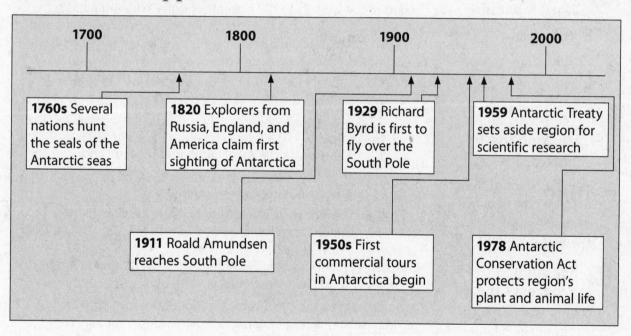

1700	1800	1900	2000

1760s Several nations hunt the seals of the Antarctic seas

1820 Explorers from Russia, England, and America claim first sighting of Antarctica

1929 Richard Byrd is first to fly over the South Pole

1959 Antarctic Treaty sets aside region for scientific research

1911 Roald Amundsen reaches South Pole

1950s First commercial tours in Antarctica begin

1978 Antarctic Conservation Act protects region's plant and animal life

Antarctica

Lesson 2: Life in Antarctica, *continued*

? **Identifying**

1. Who were the first two explorers to reach the South Pole and from which countries did they come?

_____

_____

_____

_____

Abc **Determining Word Meaning**

2. What is a *treaty*?

_____

_____

_____

✏ **Marking the Text**

3. Highlight the text that explains how decisions are made about how Antarctica is used.

✓ **Reading Progress Check**

4. What is the purpose of the Antarctic Treaty?

_____

_____

_____

_____

_____

Sharing the Land

Guiding Question *How do people share common resources and protect unique lands?*

In 1911, explorers Roald Amundsen, a Norwegian, and Robert Scott, an Englishman, raced to be the first to reach the South Pole. Their two teams made incredible journeys across brutal land. Along the way they lost men, ponies, and sled dogs to cold and starvation. In the end, Amundsen's team reached the South Pole first. Scott's team arrived 34 days later.

The remarkable feats of Amundsen and Scott gave the world an insider's view of Antarctica for the first time. Their experiences and discoveries helped scientists and explorers who came later.

Antarctica is the only continent with no native population. Therefore, when the first humans arrived, there were no other claims to the land. In 1959, 12 countries signed the Antarctic Treaty. It said that Antarctica should only be used for scientific and research purposes. A **treaty** is an agreement or contract between two or more nations, governments, or other political groups. In later years, 33 more countries signed the treaty. The purposes of the treaty are listed in the chart below.

Purposes of the Antarctic Treaty
• Set rules for the management of the land and its resources
• Protect Antarctica's environment
• Prevent weapons testing and other military actions from being carried out there

Other agreements such as the Antarctic Conservation Act relate to protecting wildlife in the region.

Some nations still claim large sections of the continent as territories. They include Argentina, Australia, Great Britain, New Zealand, and Norway. However, not all governments recognize these claims.

Antarctica is not a nation or a country and it has no government. Issues about the continent's use and protection are decided by the Antarctic Treaty System. This is a group of representatives from 48 countries that have interests in Antarctica. Important decisions affecting Antarctica's environment, wildlife, and resident scientists are made by annual meetings of the Antarctic Treaty System.

networks

Lesson 2: Life in Antarctica, *continued*

The Scientific Continent

Guiding Question *How and why do scientists study Antarctica?*

During the summer, about 4,400 people reside in Antarctica. When the temperatures fall and the constant darkness of winter sets in, the population shrinks to about 1,100 people. This is because summer weather conditions are much less severe than winter conditions. In the winter, temperatures are so low that without protective clothing, humans can freeze to death in minutes. Ice shelves along the coasts prevent ships from landing.

Governments with scientific interests have built about 50 research stations across the continent. A **research station** is a base where scientists can live and work. Each station consists of several buildings and serves several purposes.

Purposes of a Research Station
• Living quarters for scientists and staff
• Laboratories and other research facilities
• Storage areas filled with fuel, food, and equipment

An important area of study in Antarctica is the ozone layer in Earth's atmosphere. **Ozone** is a gas that absorbs harmful ultraviolet radiation from the sun. Scientists have discovered that the layer of ozone in our planet's atmosphere has thinned. The areas around Antarctica and the Southern Ocean have the lowest ozone levels on Earth. Loss of ozone affects all life on the planet. Scientists are trying to find a way to protect and preserve it.

Scientists also study the landforms under the ice shelf. They do this with a type of technology called remote sensing. **Remote sensing** uses scientific instruments placed onboard weather balloons, airplanes, and satellites to study the planet. It allows scientists to take images and measurements of objects and places that would be impossible to reach in person.

Scientists have compared rocks and fossils found in Antarctica to those found on other continents. They have found that the Antarctic continent was once part of a huge landmass called Gondwana. Millions of years ago, Antarctica, Asia, Africa, and other continents broke away and drifted across Earth's surface.

Another major area of research is climate change. Scientists have learned that the Antarctic ice sheet is shrinking. Many people are worried that this is caused by global warming. Scientists fear that climate change will permanently damage Antarctica's environment and ecosystems.

✎ Marking the Text

5. Underline the text that explains why most scientists only stay in Antarctica during the summer.

❓ Explaining

6. Why are scientists studying the ozone layer in Antarctica?

_____

_____

_____

_____

_____

🅰 Defining

7. List three types of equipment used in *remote sensing*.

_____

_____

_____

🅰 Accessing Prior Knowledge

8. Why is remote sensing particularly useful in Antarctica?

_____

_____

_____

_____

Antarctica

Lesson 2: Life in Antarctica, *continued*

Marking the Text

9. Highlight the text that explains how Antarctica is a "global barometer."

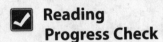
Reading Progress Check

10. Why do scientists think Antarctica was part of a larger landmass that included continents such as Africa and Asia?

_____

_____

_____

Antarctica is seen as a "global barometer." This means it is a guide to what is happening to the climate of the entire planet. Antarctica has melting ice and shrinking icebergs. This tells scientists that temperatures are rising quickly all over the globe. As icebergs melt, global sea levels will rise. This would make survival impossible in some parts of the world.

Scientists also believe that global warming is affecting the amount of plankton and krill in the region. Without enough plankton and krill to eat, fish will die off or leave the area. This leaves seals and penguins without enough fish to eat. This disruption in the local food chain could have terrible consequences for Antarctica's animal life.

About 6,000 tourists visit Antarctica every year. Many are interested in seeing the natural beauty of the area. They also want to visit research stations. However, most tourists choose Antarctica because they want an adventure more than a vacation.

Scientists studying Antarctica have helped us understand this mysterious continent. Geographers and other scientists are dedicated to their research. With each passing year they discover more clues about our planet's past. They also use what they learn to make predictions about our planet's future.

Writing
Check for Understanding

1. Narrative Describe the 1911 race to the South Pole in your own words.

_____

_____

_____

2. Informative/Explanatory What are the advantages of using remote sensing to study Earth?

_____

_____

_____

_____